The NASA STI Program Office…in Profile

Since its founding, NASA has been dedicated to the advancement of aeronautics and space science. The NASA Scientific and Technical Information (STI) Program Office plays a key part in helping NASA maintain this important role.

The NASA STI Program Office is operated by Langley Research Center, the lead center for NASA's scientific and technical information. The NASA STI Program Office provides access to the NASA STI Database, the largest collection of aeronautical and space science STI in the world. The Program Office is also NASA's institutional mechanism for disseminating the results of its research and development activities. These results are published by NASA in the NASA STI Report Series, which includes the following report types:

- TECHNICAL PUBLICATION. Reports of completed research or a major significant phase of research that present the results of NASA programs and include extensive data or theoretical analysis. Includes compilations of significant scientific and technical data and information deemed to be of continuing reference value. NASA's counterpart of peer-reviewed formal professional papers but has less stringent limitations on manuscript length and extent of graphic presentations.

- TECHNICAL MEMORANDUM. Scientific and technical findings that are preliminary or of specialized interest, e.g., quick release reports, working papers, and bibliographies that contain minimal annotation. Does not contain extensive analysis.

- CONTRACTOR REPORT. Scientific and technical findings by NASA-sponsored contractors and grantees.

- CONFERENCE PUBLICATION. Collected papers from scientific and technical conferences, symposia, seminars, or other meetings sponsored or cosponsored by NASA.

- SPECIAL PUBLICATION. Scientific, technical, or historical information from NASA programs, projects, and mission, often concerned with subjects having substantial public interest.

- TECHNICAL TRANSLATION. English-language translations of foreign scientific and technical material pertinent to NASA's mission.

Specialized services that complement the STI Program Office's diverse offerings include creating custom thesauri, building customized databases, organizing and publishing research results…even providing videos.

For more information about the NASA STI Program Office, see the following:

- Access the NASA STI Program Home Page at *http://www.sti.nasa.gov*

- E-mail your question via the Internet to help@sti.nasa.gov

- Fax your question to the NASA Access Help Desk at 301–621–0134

- Telephone the NASA Access Help Desk at 301–621–0390

- Write to:
  NASA Access Help Desk
  NASA Center for AeroSpace Information
  7121 Standard Drive
  Hanover, MD 21076–1320
  301–621–0390

NASA/TP—2004–213089

Survey of Technologies Relevant to Defense From Near-Earth Objects

R.B. Adams, R. Alexander, J. Bonometti, J. Chapman, S. Fincher, R. Hopkins,
M. Kalkstein, and T. Polsgrove
Marshall Space Flight Center, Marshall Space Flight Center, Alabama

G. Statham and S. White
ERC, Inc., Huntsville, Alabama

National Aeronautics and
Space Administration

Marshall Space Flight Center • MSFC, Alabama 35812

July 2004

Acknowledgments

The authors would like to thank Langley Research Center for supporting this worthwhile endeavor through Revolutionary Aerospace System Concepts (RASC) funding, and the Marshall Space Flight Center astronomy group for the technical assistance of Roy Young.

TRADEMARKS

Trade names and trademarks are used in this report for identification only. This usage does not constitute an official endorsement, either expressed or implied, by the National Aeronautics and Space Administration.

Available from:

NASA Center for AeroSpace Information	National Technical Information Service
7121 Standard Drive	5285 Port Royal Road
Hanover, MD 21076–1320	Springfield, VA 22161
301–621–0390	703–487–4650

EXECUTIVE SUMMARY

Nearly 300 million yr ago a comet with a diameter of ≈10 km struck the Earth near Chicxulub in present-day Mexico. The impact created a crater nearly 300 km in diameter and drove very large quantities of dust and debris into the atmosphere. The dust prevented much of the Sun's radiation from reaching the Earth's surface and the resulting Ice Age caused the extinction of ≈50 percent of the existing animal species—including, most famously, the dinosaurs. Approximately 50,000 yr ago a stony-iron asteroid with a diameter of ≈150 m struck the Earth in what is now north-central Arizona. This impact created a crater 1 mi in diameter and the resulting shockwave killed every large mammal within a radius of 24 km. In 1908, a small comet or asteroid with a diameter of ≈50 m entered the Earth's atmosphere over eastern Russia. The extreme heat and aerodynamic pressures generated during entry caused the object to disintegrate explosively at an altitude of ≈8 km. This explosive burst—since termed an airburst—occurred above Siberia, near the town of Tunguska. The airburst took place very close to the altitude at which the maximum amount of ground damage would result. It left a zone of destruction nearly 40 km in diameter around the point of disintegration.

The scientific community now accepts that these events are just major examples of the continuous ongoing bombardment of the Earth by a wide variety of objects, most of them fragments of either asteroids or comets in orbit around the Sun. Many feel that, because there appears to have been no loss of life due to cosmic bombardment, this threat can be ignored. Against this it can be argued that, as our knowledge of the solar system in general—and its minor bodies in particular—is very recent, many unexplained catastrophes in the past may actually be attributable to the impact of asteroids or comets. Additionally, the exponential growth of world population, combined with our increasing technological dependence, makes humanity much more vulnerable to the consequences of a near-Earth object (NEO) impact. Despite these worrying trends, effective means of defense against the NEO threat are available to us. That is the topic of the Marshall Space Flight Center-led study reported in this Technical Publication (TP).

This TP is divided into nine sections. Each section is intended to gradually introduce the reader to the various facets of the problem—quantifying the NEO threat, developing mitigation options, and evaluating their effectiveness. After the introduction in section 1, section 2 presents the threat of asteroids and comets, and outlines our growing understanding of NEO impacts. Section 3 discusses various mission configurations that might be selected as part of a defense strategy. Section 4 describes in detail the propulsion technologies that were considered as candidates to transport defense hardware out to an Earth-bound NEO. Section 5 reviews the actual defense or mitigation technologies that would be used to either fragment or deflect an Earth-bound NEO. Section 6 introduces the trajectory modeling tools used in the study. Section 7 describes the means by which all of the above tools were incorporated into a single "master" design environment and used to search for optimal solutions to the planetary defense problem. Section 8 contains results from parametric analyses that were conducted using the tools and data described in sections 2 through 7. Finally, conclusions and recommendations from this project are presented in section 9.

Section 2 gives an overview of our current understanding of asteroids and comets, as well as the threat that they pose to the Earth. Methods of categorization for asteroids and comets, by both orbital parameters and composition, are introduced. Next, the frequency, location, and sequence of known and suspected impacts on the Earth are presented. A timeline is then given, describing the development of human knowledge of asteroids, comets, and the impacts from these bodies. This timeline helps to explain why our historical record does not contain more information on possible impacts. Current knowledge of the NEO population within the solar system is then discussed. The immediate physical effects of various types of impact are then described, as are the longer-term consequences. Finally, some thoughts are offered on the current lack of significant public credibility regarding this threat.

There are many different options that are potentially available to defend against an incoming NEO. Grouping these options into categories allows us to develop analysis processes applicable across an entire category, rather than having to establish a different process for each option. This categorization approach is new to this study; evidence of its previous use was nowhere in the literature.

At the highest level, mitigation options can be divided into two categories: deflection and fragmentation. The deflection option leaves the object largely intact, but changes its velocity by a small amount, sufficient to ensure that it will miss the Earth by a distance greater than or equal to a certain minimum value—in the literature, this is usually set at 3 Earth radii. The fragmentation option breaks the NEO into pieces, each small enough to burn up in the Earth's atmosphere. There is some uncertainty over the maximum fragment size that would not threaten the Earth; a diameter of 10 m was selected here as a first approximation. The means by which energy is delivered—to either fragment or deflect an NEO—can also be categorized. A remote station would project a beam or fire a projectile at the incoming NEO. An interceptor would travel across the inner solar system and actually impact on the NEO. Finally, under the rendezvous option, the mitigation hardware would be transported out to the NEO and would match orbits with it. This allows the vehicle to operate on or near to the NEO for an extended duration, delivering its deflection or fragmentation energy over an extended time instead of in one short impulse.

Several propulsion systems were considered as candidates to place the mitigation vehicle onto either an intercept or a rendezvous trajectory with the NEO. A multistage liquid oxygen/liquid hydrogen rocket was selected as the baseline option due to its technical maturity. The nuclear thermal rocket was retained as an additional option because of its superior specific impulse. A scaled-down derivative of the ORION concept of the 1960's was also retained—referred to as the nuclear pulse technique. For trajectories in the inner solar system requiring large inclination changes, solar sails are highly competitive and so were likewise considered. Finally, a derivative of the solar sail—the solar collector—was considered.

In addition, several very interesting concepts for threat mitigation were considered. Fragmentation of the incoming NEO using nuclear devices was assessed. The use of nuclear devices to deflect the NEO was also considered. The solar sail was considered, but was rejected, as it was found to be impractical for all but the smallest asteroids. The solar collector, however, showed remarkable promise for all but the largest asteroids and comets. The novel option of deflection by use of a rapidly growing, pulsed magnetic field was considered. Furthermore, the use of a mass driver, transported out to and installed on the incoming asteroid, was considered and was modeled in some detail. Finally, deflection of the NEO by purely kinetic means—utilizing an ultra-high velocity inert projectile—was considered.

Both the outbound and inbound trajectories were modeled using tools developed under this study. The outbound model solved the Gauss problem for high thrust trajectories. The inbound model calculated backwards in time from the point of impact with the Earth in order to determine the instantaneous ΔV necessary to achieve the minimum required deflection. The inbound trajectory calculations accounted for the influence of both the Sun and the Earth. Inclusion of the Sun's gravitational field in these calculations yields significantly different results than the two body approximations found in the literature. Some of our propulsive techniques do not provide high thrust levels and, as they neglect gravity losses, these trajectory calculation techniques are only approximate. However, for this first attempt at a solution, we were willing to accept the consequent level of uncertainty in our answers. We intend to revisit this issue and model the effects of gravitational losses at the earliest opportunity in the future.

Our threat parametric builds on the results of previous studies in modeling the overall consequences of an impact. Our model uses the current knowledge of the asteroid and comet population to execute Monte Carlo simulations to establish the probability of impact. These results are combined with estimates of the average number of fatalities for an asteroid or comet of a given size and composition and are then used to determine the average number of deaths resulting from this threat over a given time period. Our various concepts for threat mitigation can then be evaluated by the percentage of this threat that each can defeat, and ultimately, by the mean number of lives that would be saved.

The next objective was to combine the tools and concepts described above to determine the optimal configurations to defeat the threat. Time and funding constraints necessitated a less ambitious approach. Assuming a baseline asteroid orbit, we integrated our trajectory, propulsion, and threat mitigation tools to quantify the relationship between required system mass and size of object deflected. These results are summarized for the concepts evaluated in the study in the table below. Based on these results, as well as qualitative comparisons documented here, we considered the nuclear pulse option the most viable for the overall threat. However, recognizing our limited modeling capability for this study, we strongly recommend a broad spectrum of deflection technologies, including all options evaluated here, be considered for any future work.

System	Maneuver	Time Before Impact (days)/Outbound Travel Time (days)	Total System Mass at SOI (t) for Different Asteroid Diameters (m)			Maximum Diameter of Asteroid*(m)/ Total System Mass at Earth SOI (t)
			100	1,000	10,000	
Staged chemical/ mass driver	Rendezvous	2,900/2,400	NA	NA	NA	50/6,849 80/6,918
Staged chemical/ nuclear deflection	Intercept	1,509/910	0.847	8.27	1,300	9,000/1,000
	Rendezvous	1,075/132	5.62	568	87,800	1,000/1,000
Staged chemical/ kinetic deflection	Intercept	1,025/800	73.8	NA	NA	260/1,000
Nuclear pulse	Rendezvous	2,170/1,200	29.7	41.8	1,240	9,000/1,000
Solar collector	Rendezvous (≈3 yr)	1,076/65**	0.637	1.07	167	§
	Rendezvous (≈10 yr)	3,635/115**	0.550	0.636	34.6	§

 * Maximum was constrained to a total system mass at Earth SOI of 1,000 t.
 ** Times are for 100-m-diameter chondrite. Rendezvous times are greater for larger asteroids, although total missions times change little.
 § The solar collector system is limited more by solar collector size than by total system mass. See figure 112.

TABLE OF CONTENTS

1. INTRODUCTION 1

2. THREAT 3

    2.1 Categorization of Asteroids and Comets 3
    2.2 Earth's Impact Record 7
    2.3 History Related to Near-Earth Objects 13
    2.4 Measuring the Near-Earth Object Population 16
    2.5 Damage Mechanisms 18
    2.6 The Credibility Problem 25
    2.7 The Torino Scale 29

3. MISSION CONFIGURATIONS 31

    3.1 Deflection Versus Fragmentation 31
    3.2 Remote Station Versus Interception Versus Rendezvous 33

4. OUTBOUND PROPULSION 37

    4.1 Staged Chemical 38
    4.2 Nuclear Thermal Rocket 43
    4.3 Nuclear Pulse 46
    4.4 Solar Sail 53
    4.5 Solar Collector 60

5. THREAT MITIGATION 62

    5.1 Nuclear Fragmentation 62
    5.2 Nuclear Deflection 66
    5.3 Solar Sails 71
    5.4 Solar Collector 74
    5.5 Magnetic Flux Compression 76
    5.6 Mass Driver 81
    5.7 Kinetic Deflection 93

6. TRAJECTORY MODELING 102

    6.1 Outbound 102
    6.2 Inbound 103

TABLE OF CONTENTS (Continued)

7. THREAT PARAMETRIC	110
8. PARAMETRIC RESULTS	115
8.1 Integrated Analysis	115
8.2 Architecture Options	117
8.3 Parametric Performance	118
9. CONCLUSIONS AND RECOMMENDATIONS	144
9.1 Public Awareness	144
9.2 Statistical Problem	144
9.3 Funding of Future Work	145
9.4 Development and Deployment of Mitigation Systems	145
9.5 Accomplishments	146
9.6 Assessment of Mitigation Options	147
9.7 Future Work	148
9.8 Summary Conclusion	149
APPENDIX A—CURRENT NEAR-EARTH OBJECT SEARCH PROGRAMS	150
A.1 SpaceWatch	150
A.2 Spaceguard	150
A.3 Lincoln Near-Earth Asteroid Research	150
A.4 Near-Earth Asteroid Tracking	151
A.5 Lowell Observatory Near-Earth Object Search	151
APPENDIX B—SOLAR ARRAY CALCULATIONS	152
B.1 Method I	152
B.2 Method II	152
APPENDIX C—MASS DRIVER	154
C.1 Model of the Forces on a Bucket Coil Due to the Nearby Drive Coils	154
C.2 Drive and Bucket Coil Currents	158
C.3 Analysis of Bucket Kinetic Energy and Acceleration	161
C.4 Drive Coils	163
C.5 Braking Coils	166
C.6 Bucket Design	166
C.7 Interference Between Adjacent Drive Coils	170
C.8 Effect of Bucket Coil Motion on Stationary Coil Circuit Operation	176

TABLE OF CONTENTS (Continued)

C.9 Drive Coil	180
C.10 Braking Coil	181
C.11 Self-Induced Coil Stresses	182
APPENDIX D—COIL FORCE MODEL	185
REFERENCES	190

LIST OF FIGURES

1. Organization of RASC FY 2002 activities 1

2. Orbits of 2062 Aten, 1862 Apollo, and 1221 Amor relative to Earth and Mars 4

3. Orbits of representative short-period comets relative to the outer planets. Halley's Comet represents the Halley class and Comet Crommelin represents the Jupiter class 5

4. Orbits of representative short- and long-period comets relative to the solar system. Comet Hale-Bopp represents long-period comets 6

5. Location of known impact craters noting diameters as of 1998. Age distribution of these craters is also included 8

6. Projected area affected from the Tunguska blast of 1908. Arrows depict the location and direction trees were knocked down from the blast 9

7. Tunguska impact area superimposed over Madison County, Alabama, USA. Several hundred thousand casualties can be expected from such an impact 10

8. Aerial view of the Barringer Impact Crater in Arizona, USA 11

9. Calculated impact areas from the Barringer meteor 12

10. Extinction events versus time according to the fossil record 13

11. Location of known minor planets on March 2, 2002, plotted relative to the inner planets. NEOs are red, other asteroids are green, and comets are blue 16

12. Number of known NEAs versus time. Note the rapid increase in discoveries in recent years due to the use of CCDs and increased interest in the asteroid and comet threat 17

13. Deep-water wave height at 1,000 km distance versus initial meteor radius for soft stone meteor 20

14. Deep-water wave height at 1,000 km distance versus initial meteor radius for Fe meteor 20

LIST OF FIGURES (Continued)

15.	Blast wave damage versus impact energy	21
16.	Density and optical depth of atmospheric dust versus impact energy	23
17.	Blast wave damage versus impact energy	24
18.	Mass of water lifted into the atmosphere versus impact energy	25
19.	Predicted world population in the last century. Population is extrapolated through the middle of this century	27
20.	Human population density graph for all continents (except Antarctica). Greenland and Iceland are not represented as well as some Pacific islands	28
21.	Illustration of the various category of threat under the Torino Scale	29
22.	Categories within the Torino Scale	30
23.	Illustration of deflection method of threat mitigation	32
24.	Illustration of fragmentation method of threat mitigation	32
25.	Delivering deflection or fragmentation energy by the remote station mode	34
26.	Delivering deflection or fragmentation energy by the interception mode	34
27.	Delivering deflection or fragmentation energy by the rendezvous mode	35
28.	Two-stage lox/LH$_2$ vehicle. Image produced by INTROS	38
29.	Regression curve fit of lox/LH$_2$ launch vehicle stages	39
30.	Schematic of a nuclear thermal rocket	43
31.	Regression curve fit of NERVA program-developed engines	44
32.	U.S. Air Force ORION spacecraft (1964)	47
33.	NASA Gabriel spacecraft (1999)	47
34.	EPPP concept vehicle	50

LIST OF FIGURES (Continued)

35.	Shock absorber operations	51
36.	Artists concept of a billowing, square solar sail	54
37.	Schematic and dimensions for a square solar sail	54
38.	Sail trajectories relative to lightness number	58
39.	Solar collector configuration	60
40.	Blast yield—explosive placed at center of body—required for fragmentation as a function of asteroid radius	63
41.	Device mass versus explosive yield	64
42.	Geometric position of the pulse unit to the planetary body and cone half-angle definition	69
43.	Nuclear pulse rocket delivery system sketch	70
44.	Deflection ΔV imposed on 10-m-diameter asteroid	72
45.	Deflection ΔV imposed by 6 gm/m^2 solar sail with varying areas on varying diameter asteroids	73
46.	Schematic of solar collector	74
47.	ΔV imposed on varying asteroid sizes by 100-m-diameter solar collector	75
48.	Critical disk diameter	78
49.	Magnetic flux compression generato	80
50.	E-bomb magnetic flux compression generator	80
51.	Coil gun conceptual design	83
52.	Relative current flow directions of bucket and immediately adjacent drive coils	83
53.	Relative current flow directions of bucket and nearest drive coils	84
54.	Relative current flow directions of bucket and surrounding drive coils	85

LIST OF FIGURES (Continued)

55.	Relative current flow directions of bucket and nearby drive coils	85
56.	Mass driver operation	86
57.	Relative current flow directions of bucket and braking coils	86
58.	Mass driver system schematic	87
59.	Mass driver system view	89
60.	Main components of the mass driver	89
61.	System following discharge—bucket returns for reloading	90
62.	Bucket about to be reloaded and cooled prior to next discharge	90
63.	Loaded bucket about to enter mass driver	91
64.	Bucket containing expellant under acceleration within mass driver	91
65.	Discharge in progress—bucket is decelerating while expellant mass exits mass driver at high speed	92
66.	Following discharge—decelerated bucket exits mass driver and joins return system	92
67.	Interception geometry	94
68.	Impact and ejection geometry	95
69.	Fragmentation data and curve fit results	100
70.	Outbound trajectory ΔVs for 3,600-day total mission duration	103
71.	Illustration showing a typical NEO orbit. The velocity of the planetary body at impact for this case is $(-40,0,0)^T$ km/s, parallel to the x axis of the Heliocentric-Ecliptic system	104
72.	M1999JT6 orbit plot	105
73.	Required impulsive ΔV for 42 km/s velocity for various maneuvers to avoid collision with Earth, showing the benefit of the UP maneuver when impact is only a few days away	107

LIST OF FIGURES (Continued)

74. Required impulsive ΔV for 42 km/s velocity for various maneuvers to avoid collision with Earth, showing the benefit of the DECEL and OUTSIDE maneuvers when impact is several weeks away 107

75. Required impulsive ΔV for 35 km/s velocity for various maneuvers to avoid collision with Earth, showing the benefit of the DECEL maneuver when impact is only a few days away 108

76. Required impulsive ΔV for various maneuvers to avoid collision with Earth for planetary body with velocity of 35 km/s 108

77. Required impulsive ΔV for various maneuvers to avoid collision with Earth for planetary body with velocity of 35 km/s (long lead time) 109

78. Average deaths from single asteroid impact versus size, and the average total number of deaths prevented if all impacts of equal or less energy can be avoided 114

79. Proposed analysis process for assessing total amount of threat mitigated 116

80. Staged chemical/mass driver model 119

81. (a) Optimal deflection direction, (b) optimal deflection direction—detailed view, and (c) optimal deflection direction—detailed view—minimal deflection ΔV 120

82. Variation of mass driver total system mass with required asteroid deflection ΔV for a 50-m-diameter chondrite 121

83. Staged chemical/mass driver vehicle mass at Earth departure 122

84. Staged chemical/mass driver vehicle mass at Earth departure (expanded view) ... 123

85. Optimal staged chemical/mass driver mission 123

86. Staged chemical/mass driver vehicle mass versus chondrite asteroid diameter 124

87. Mass driver deployed mass and total operating time versus chondrite asteroid diameter (50–100 m) 124

88. Mass driver deployed mass and total operating time versus chondrite asteroid diameter (100–1,000 m) 125

LIST OF FIGURES (Continued)

89.	Diagram of the ModelCenter setup for the staged chemical/nuclear deflection option	125
90.	Total system mass for the staged chemical/nuclear blast option versus total mission time for various rendezvous times. Here, the staged chemical system does not match the asteroid's orbit at encounter.	126
91.	Total system mass for the staged chemical/nuclear blast option versus total mission time for various rendezvous times (zoomed). Here, the staged chemical system matches the asteroid's orbit at encounter.	127
92.	Minimum total system mass for the staged chemical/nuclear blast option, showing the optimum rendezvous and total mission times for intercept	127
93.	Minimum total system mass for the staged chemical/nuclear blast option, showing the optimum rendezvous and total mission times for rendezvous	128
94.	Optimum intercept trajectory for the staged chemical/nuclear deflection option	128
95.	Optimum intercept trajectory for the staged chemical/nuclear deflection option	129
96.	Minimum total system mass for the staged chemical/nuclear blast option versus chondrite diameter for both intercept and rendezvous	130
97.	Minimum total system mass for the staged chemical/nuclear blast option versus chondrite diameter for the smaller chondrites, showing both intercept and rendezvous	130
98.	Staged chemical/kinetic deflection model	131
99.	(a) Staged chemical/kinetic deflection vehicle mass at Earth departure, (b) detailed view, and (c) detailed view—minimum mass solution	133
100.	Optimal staged chemical/kinetic deflection mission	134
101.	Staged chemical/kinetic deflection vehicle mass versus chondrite asteroid diameter	135
102.	Projectile mass versus chondrite asteroid diameter	135
103.	Diagram of the ModelCenter setup for the nuclear pulse option	136
104.	Minimum total system mass for the nuclear pulse option, showing the optimum rendezvous and total mission times	137
105.	Optimum rendezvous trajectory for the nuclear pulse option	137

LIST OF FIGURES (Continued)

106.	Minimum total system mass for the nuclear pulse option versus chondrite diameter	138
107.	Minimum total system mass for the nuclear pulse option versus chondrite diameter for the smaller chondrites	138
108.	Plot of the difference between required outbound and inbound solar collector sizes. Negative values mean that the inbound requirement dominates.	139
109.	Combinations of total mission time and rendezvous time where inbound and outbound required solar sail sizes are equal, and the associated total system mass for 500-m-diameter chondrite	140
110.	Diagram of the ModelCenter setup for the solar collector option	141
111.	Optimum rendezvous trajectory for the solar collector option for a 100-m-diameter chondrite	142
112.	Minimum total system mass and size for the solar collector option versus chondrite diameter	142
113.	Minimum total system mass and size for the solar collector option versus chondrite diameter for the smaller chondrites	143
114.	Drive coils included in and omitted from the analysis	154
115.	Drive coils included in the analysis	155
116.	Elliptical function curve fits—$K(m)$ and $E(m)$ versus m	157
117.	Bucket coil current directions	159
118.	Bucket and drive coil current directions during acceleration	160
119.	Ideal current versus time profile for a single drive coil	163
120.	Sinusoidal current versus time profile for a single drive coil	164
121.	Oscillating drive circuit—charging capacitor	164
122.	Oscillating drive circuit—discharging capacitor	164
123.	Real current versus time profile for a single drive coil	165

LIST OF FIGURES (Continued)

124.	Bucket conceptual design	166
125.	Bucket handling through a complete cycle	168
126.	Discharge of isolated drive coil	170
127.	Current flow directions of two adjacent drive coils	172
128.	Simple model of a drive coil circuit	176
129.	The effect of bucket acceleration	181
130.	Self-induced magnetic field and resulting force	182

LIST OF TABLES

1. Description of asteroid and comet compositions and representative predicted densities 6
2. Comparable terrestrial events for NEOs of various diameters 7
3. Timeline of scientific discoveries relevant to humanity's knowledge of asteroids and comets 15
4. Recent near misses by comets and asteroids. By comparison, the distance between the Earth and the Moon is ≈240,000 mi. 18
5. Consequences of impact by NEOs of various sizes 26
6. Data on in-service and historical launch vehicles that have lox-/LH$_2$-powered stages 40
7. Qualitative considerations for outbound propulsion using chemically-powered rockets 42
8. Data on nuclear rocket engines developed under the NERVA program 44
9. Qualitative considerations for outbound propulsion using nuclear thermal rockets 45
10. Qualitative considerations for outbound propulsion using nuclear pulse 48
11. Optical, billowing force and other parameters used in solar sail analysis 56
12. Qualitative considerations for outbound propulsion using solar sails 59
13. Qualitative considerations for outbound propulsion using solar collectors 61
14. Nuclear device masses 64
15. BLU–113 penetrator characteristics 65
16. Qualitative considerations for threat mitigation using nuclear fragmentation 65
17. Qualitative considerations for threat mitigation using nuclear deflection 67
18. Chemical rocket assumptions 71

LIST OF TABLES (Continued)

19.	Qualitative considerations for threat mitigation using a solar sail	73
20.	Qualitative considerations for threat mitigation using a solar collector	76
21.	Qualitative considerations for threat mitigation using magnetic flux compression	81
22.	Qualitative considerations for threat mitigation using the mass driver	93
23.	Relative size of largest fragment at various collisional energies	99
24.	Qualitative considerations for threat mitigation using kinetic deflection	101
25.	Original and modified orbital elements of 1999JT6	104
26.	Explanation of the different maneuvers available for use in the program PBI	106
27.	Types of planetary bodies examined in the Monte Carlo simulation and their average contribution to the total number of deaths over the next century	110
28.	Architecture options considered in this study	118
29.	Causes of death and associated probabilities for a U.S. resident	145
30.	Summation of parametric results for mitigation concepts	147

LIST OF ACRONYMS AND SYMBOLS

AANEAS	Anglo-Australian Near-Earth Asteroid Survey
au	astronomical unit
carb	carbonaceous chondrite
CCD	charged-coupled device
CF	catastrophic fragmentation
DOD	Department of Defense
EMP	electromagnetic pulse
EPPP	external pulsed plasma propulsion
Fe	iron
GEODSS	ground-based electro-optical deep space surveillance
GMD	global missile defense
H	hardness parameter for asteroidal material
H_2O	water
HOPE	human outer planet exploration
ICBM	intercontinental ballistic missile
ICR	inductance capacitance resistor
JPL	Jet Propulsion Laboratory
K-T	Cretaceous-Tertiary (extinction event)
LaRC	Langley Research Center
LCR	inductance-capacitance-resistance

LIST OF ACRONYMS AND SYMBOLS (Continued)

LEO	low-Earth orbit
LH_2	liquid hydrogen
LINEAR	Lincoln near-Earth asteroid research
LN_2	liquid nitrogen
LONEOS	Lowell Observatory near-Earth object search
lox	liquid oxygen
LP	long period (comet)
Ma	mega annum (million years)
MSFC	Marshall Space Flight Center
MSSS	Maui Space Surveillance Site
MTM	momentum transfer mechanism
NEA	near-Earth asteroid
NEAT	near-Earth asteroid tracking
NEO	near-Earth object
NERVA	nuclear engine for rocket vehicle applications
Ni	nickel
NO	nitrogen oxide
N_2O_4	nitrogen tetroxide
NTR	nuclear thermal rocket
PBI	planetary body intercept
PBM	planetary body maneuvering

LIST OF ACRONYMS AND SYMBOLS (Continued)

PBO	planetary body—outbound
PHA	potentially hazardous asteroid
RASC	revolutionary aerospace systems concept
SOI	sphere of influence
SP	short period (comet)
TD30	Advanced Concepts Department
TNT	tri-nitro-toluene
TP	Technical Publication
UDMH	unsymmetrical dimethyl hydrazine
USA	United States of America

NOMENCLATURE

A	area—solar array, solar sail, or cross sectional
a	semimajor axis of orbit (astronomical units or kilometers)
a_0	solar sail characteristic acceleration
a_B	radius of bucket coil
a_D	radius of drive coil
B	magnetic flux density; magnetic field
B_b	non-Lambertian coefficient for backside of sail
B_f	non-Lambertian coefficient for front side of sail
C	electrical capacitance
C_1, C_2, C_3	solar sail force coefficients
C_{LN_2}	specific heat of liquid nitrogen
c	speed of light
D	spherical diameter
$D_{standoff}$	standoff distance
E	elliptical integrals of the second kind; electric field; energy
E_P	energy of collision
e	orbital eccentricity
F	number of expected average fatalities per year for all impacts of some given energy or less; total radial force acting on a current-carrying coil; solar flux; force; thrust
F_{an}	absorptive force normal to solar sail centerline
F_{at}	absorptive force tangential to solar sail centerline
F_E	effectivity factor
F_e	emissive force tangential to solar sail centerline
F_{inert}	inert mass fraction
F_n	solar sail force normal to solar sail centerline

NOMENCLATURE (Continued)

F_{norm}	total force adjusted for billowing
F_{rn}	reflective force normal to solar sail centerline
F_{rt}	reflective force tangential to solar sail centerline
F_t	solar sail force tangential to solar sail centerline
f	number of expected average fatalities per year for an impact of a given energy; radial force per unit length of coil
f_i	substitute variable (used in bucket coil motion analysis); $i = 1,2,3,4$
G	constant of gravitation
g_0	gravitational constant at Earth's surface
H	asteroid "hardness"
H_c	total thermal capacity of bucket coils and their LN_2 coolant
H_{vap}	heat of vaporization
h	bucket internal height
I	orbital inclination (degrees); electrical current
I_0	initial current
I_B	current flowing in bucket coil; solar sail front side emission coefficient
I_D	current flowing in drive coil
I_{D0}	maximum current flowing in drive coil
I_j	current flowing in coil j
$I(r)$	luminous intensity
I_{sp}	specific impulse
I_{sp_n}	specific impulse of the nth stage
J	current density
K	elliptical integrals of the first kind; thermal conduction constant between bucket LN_2 reservoir and external cold plate
KE	kinetic energy
k	variable used to evaluate elliptical integrals
L	electrical self-inductance; conductor length

NOMENCLATURE (Continued)

L_s	solar luminosity
l	diameter of solar collector or curved sail; length; orbital inclination (degrees)
M	mutual inductance between two current-carrying, single-turn coils; mass of solar array; magnification factor for solar collector
M_B	mass of loaded bucket
M_{body}	mass of body
M_{ej}	mass of material ejected due to the interceptor impact
M_f	final mass
M_{f_n}	final mass of nth stage
M_i	mass of interceptor; initial mass
M_{ij}	mutual inductance between bucket coils i and j
M_{i_n}	initial mass of nth stage
$M_i(x)$	mutual inductance between one turn of a bucket coil and one turn of drive coil I when the bucket coil is a distance x from the drive coil
M_L	mass of largest fragment
M_{LN_2}	mass of liquid nitrogen used to cool the bucket superconducting coils
M_{NEO}	mass of near-Earth object
M_n	mutual inductance between one turn of a bucket coil and one turn of stationary coil n
MR_n	mass ratio of nth stage
M_S	mass of Sol (the Sun)
M_T	mass of target NEO
m	mass
$\dot{m}_e$	mass flow rate of ejecta
m_{pay}	payload mass
m_{pay_n}	payload mass of nth stage
m_{p_n}	propellant mass of nth stage
m_{reac}	NTR reactor mass
m/s	mass relative to Earth

NOMENCLATURE (Continued)

m_{s_n}	inert mass of nth stage
N	number of circuits in an electrical system; number of turns in coil
N_B	number of turns per bucket coil
N_D	number of turns per drive coil
N_{DC}	total number of drive coils
N_i	number of turns in coil i
P	electrical power
P_{ej}	momentum of ejected material
P_{jet}	jet power
p_o	total pressure
Q, q	electrical charge
R	resistance of coil; radius of coil
R_a	asteroid orbital radius
R_c	crater radius
R_E	Earth's orbital radius
R_S	radius of Sun
r	bucket internal radius; radius of vehicle from Sun; distance from coil to the target
r'	reflection coefficient; idealized radius of target body
$\ddot{r}$	acceleration of orbiting body
r_o	wire radius
$\mathbf{r}$	position vector
$\mathbf{r}_{E/S}$	position vector of the Earth relative to the Sun (other objects use similar notation; time derivatives use the overdot notation)
S	distance between adjacent stationary coils; standard intercoil distance
s'	spectral reflection coefficient
T	temperature; tension
T_{cp}	temperature of cold place
T_H	maximum temperature of liquid nitrogen used to cool bucket superconducting coils

NOMENCLATURE (Continued)

T_L	minimum temperature of liquid nitrogen used to cool bucket superconducting coils
t	time
U	total magnetic energy
U_e	total electrostatic energy
U_M	magnetic energy
V	relative speed of collision; bucket internal volume; electrical potential
$\mathbf{V}_{E/S}$	velocity vector of the Earth relative to the Sun (other objects use similar notation)
v	velocity
v_e	velocity of ejecta
v_i	interceptor speed; mass driver bucket speed as it passes drive coil i
v_{min}	minimum speed of ejected material
W	required blast yield
x	distance between bucket and drive coils
Y	NEO material strength
α	Lagrange multiplier; angle of attack
β	sail lightness number
β'	gas expansion factor
ΔKE	change in kinetic energy experienced by mass driver bucket as it moves between two adjacent drive coils
ΔQ	generated in mass driver bucket during one operational cycle; total amount of heat; total amount of heat generated in mass driver bucket during one operational cycle
δ	skin depth
δp	coefficient, vertical component of momentum
ε	overall efficiency of solar electric power system; induced electromagnetic field
ε_b	emission coefficient for back
ε_f	emission coefficient for front
ε_n	inert mass fraction

NOMENCLATURE (Continued)

η	ratio of bucket internal height to internal radius; efficiency of solar sail
λ	variable used in equation solutions
λ_n	payload mass fraction for nth stage
μ_0	permeability of free space
ν	ejecta speed; final mass driver bucket speed mass driver stationary coil current oscillation frequency
θ	solar sail cone angle
φ	variable of integration used to evaluate elliptical intgrals; angle between impact and ejecta
ρ	NEO material mass density; resistivity
σ	solar array mass per unit area or solar sail loading parameter; conductivity
σ'	Stefan-Boltzmann constant
τ	time during which bucket must remain in contact with external cold plate in order for its coils to be cooled after one operational cycle
Φ	magnetic flux
Φ_i	total magnetic flux through bucket coil i
ϕ	solar sail centerline angle
Ω	longitude of the ascending node (degrees)
ω	argument of Perihelion (degrees); oscillation frequency of coil current
$\vec{\ell}$	coil path vector

Subscripts

E	denoting the Earth
PB	denoting the planetary body
S	denoting the Sun

TECHNICAL PUBLICATION

SURVEY OF TECHNOLOGIES RELEVANT TO DEFENSE FROM NEAR-EARTH OBJECTS

1. INTRODUCTION

In FY 2002, the revolutionary aerospace systems concepts (RASCs) activity, managed from Langley Research Center (LaRC) selected a broad range of projects for the year's activities. These projects were organized into five groups as shown in figure 1. Marshall Space Flight Center (MSFC) participated in two of these groups, led from TD30/Advanced Concepts. The work completed for group 2—human outer planet exploration (HOPE)—is documented in *NASA/TP—2003–212691*.[1] MSFC's participation in group 4 activities is documented in this Technical Publication (TP). Several projects were funded under group 4; however, MSFC's activities were confined to planetary body maneuvering (PBM).

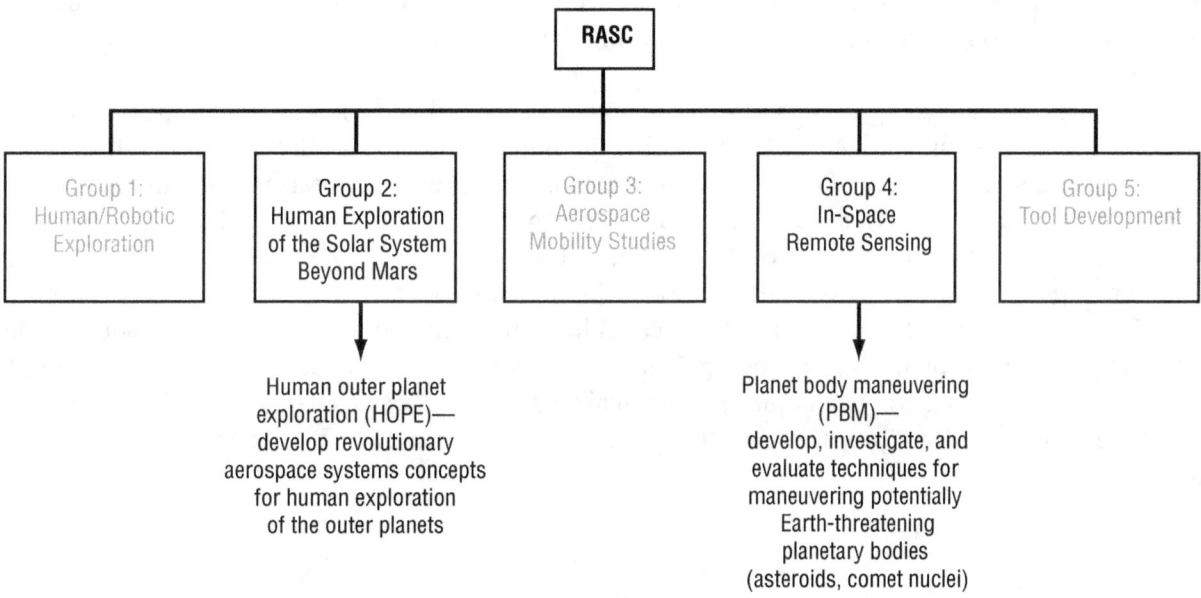

Figure 1. Organization of RASC FY 2002 activities.

Work under the PBM project was confined to defense of the Earth from collisions from asteroids and comets. Many of the technologies developed for protective maneuvering of planetary bodies are also applicable to maneuvering these bodies for resource utilization. Asteroids and comets can be maneuvered—in a careful and controlled way—close to the Earth to be mined for structural materials and water. The mass of these bodies could also anchor rotating tethers and skyhooks. Finally, such bodies could be intentionally targeted to impact Mars or Venus in order to alter rotation speed and/or atmospheric composition.

Despite these other potential applications for maneuvering technologies, this project concentrated solely on planetary defense as the most critical mission—a uniquely suitable one for NASA. This can be seen immediately from the NASA mission statement:

To understand and protect our home planet
To explore the Universe and search for life
To inspire the next generation of explorers
… as only NASA can.

It has been suggested that the mission of planetary defense is best suited to the Department of Defense (DOD). This argument is based on the DOD's extensive experience in the interception of high-speed objects. However, the very high energies necessary for deflection of massive planetary bodies combined with the unique problems of operation in interplanetary space suggest that NASA will have a major, if not a leading, role to play. The above mission statement suggests very strongly that the Agency should address this threat. NASA's unique capabilities may well make it the most uniquely qualified organization in the world to take on the daunting task of protecting the planet from this threat.

Research conducted by the TD30 PBM team to understand and categorize the threat of impact by an asteroid or comet is summarized in this TP. Using the limited knowledge currently available on the solar system's asteroid and comet population, an analytical tool was developed to estimate the number of human lives that could potentially be lost because of this threat over a specified period of time. Propulsion technologies suitable for reaching the approaching object were then researched and deflection methods investigated. Analytical tools were developed to model the actual deflection techniques. These various tools were then linked with an additional set of tools capable of modeling both inbound and outbound trajectories. Parametric results could then be generated using the linked propulsion, deflection, and trajectory tools to calculate optimal deflection techniques for use against specific threat scenarios. Finally, these parametric results are presented and a set of conclusions established as to the effectiveness of each deflection method.

2. THREAT

2.1 Categorization of Asteroids and Comets

Asteroids and comets are both categorized as minor planets. This designation is appropriate on the grounds of both size and orbital parameters. An additional level of categorization can also be established on the basis of composition. These categorizations are discussed further below.

Distinguishing between asteroids and comets is not entirely straightforward. Asteroids are small objects (<1,000-km diameter), usually in eccentric, low-inclination solar orbits. The majority of asteroids have both aphelia and perihelia between the Martian and Jovian orbits, although there are exceptions. They are thought to be largely of rocky and/or metallic composition.

By comparison, comets can be distinguished from asteroids by both orbit and composition. Cometary orbital eccentricities are usually higher than those of asteroids ($e > 0.35$ for most comets, with some even having $e > 0.9$). Cometary aphelia are usually located beyond the orbits of the gas giants. Comets are ice-rich bodies that become visually prominent when heat from the Sun causes their trapped volatiles to sublimate. The most visible and distinctive features of comets are the coma and tail, produced by the release of these volatile compounds, and also by dust. Most of the mass of a comet is contained within a comparatively tiny central nucleus, now thought likely to have been formed directly from the primordial solar nebula.

Asteroids can also be categorized according to their orbital parameters. Each category is usually named after its first representative. Hence, the asteroids 2062 Aten, 1862 Apollo, and 1221 Amor are all significant because they name their respective categories. Asteroids in the Aten class have orbits with apehelia <0.983 au and semimajor axes less than that of the Earth. Apollo asteroids have semimajor axes greater than Earth's (<1 au) and perihelia <1.017 au. Finally, Amors have orbits between Earth and Mars (1 au $< r <$ 1.3 au). Even though these orbits do not necessarily cross that of the Earth, asteroids in this class could easily be perturbed into a collision trajectory. The orbits of Aten, Apollo, and Amor class asteroids are illustrated in figure 2.

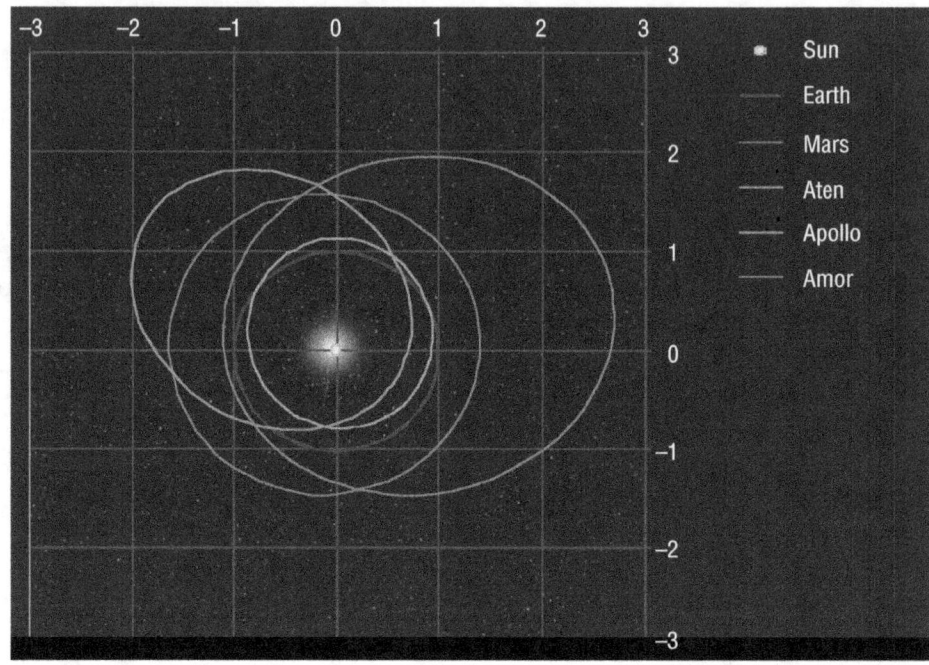

Figure 2. Orbits of 2062 Aten, 1862 Apollo, and 1221 Amor relative to Earth and Mars.

Their orbits also categorize comets. Comets are generally listed as short or long period. Short-period comets are further broken down into Halley and Jupiter classes. Jupiter class comets have perihelia within Jupiter's orbit. Although they can have longer periods, clearly shown by the 76-yr period of the example from which they take their name, Halley's comets remain within the solar system. The orbits of Halley's Comet and a representative Jupiter class comet are shown in figure 3. Short-period comets are believed to originate from the Kuiper belt and generally have inclinations <30°. Additionally, their orbital periods are <200 yr.

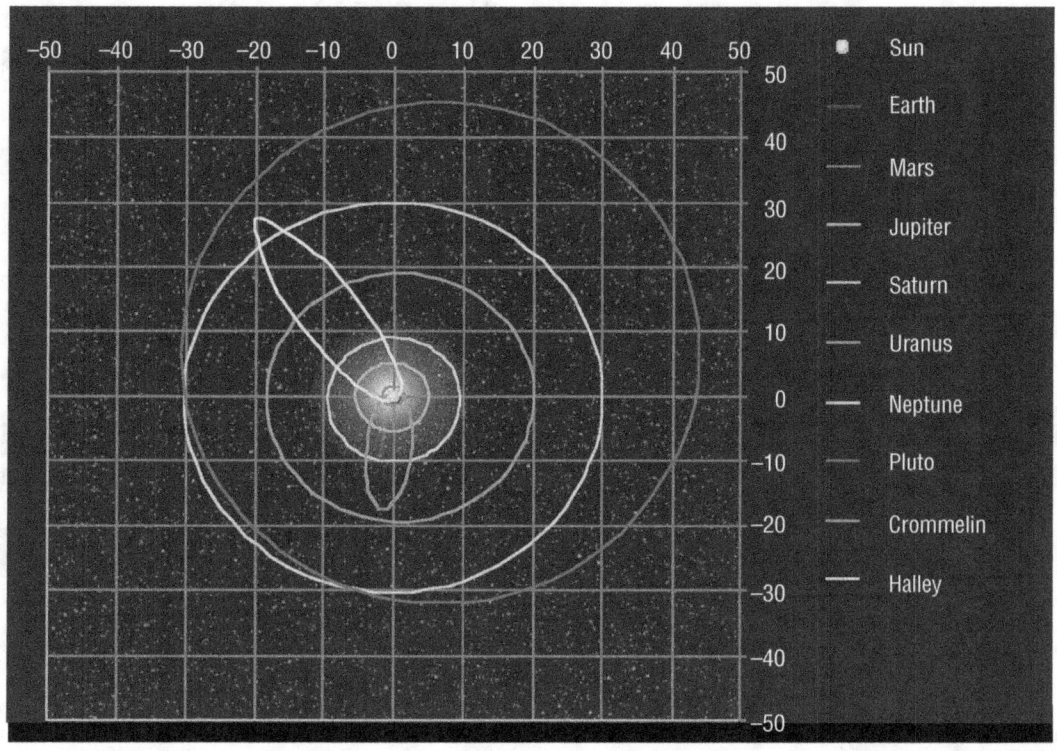

Figure 3. Orbits of representative short-period comets relative to the outer planets. Halley's Comet represents the Halley class and Comet Crommelin represents the Jupiter class.

Long-period comets are believed to come from the Oort cloud; they have periods well above 200 yr and appear to have no preferred orbital inclination. The orbit of a representative long-period comet, Hale-Bopp, is found in figure 4.

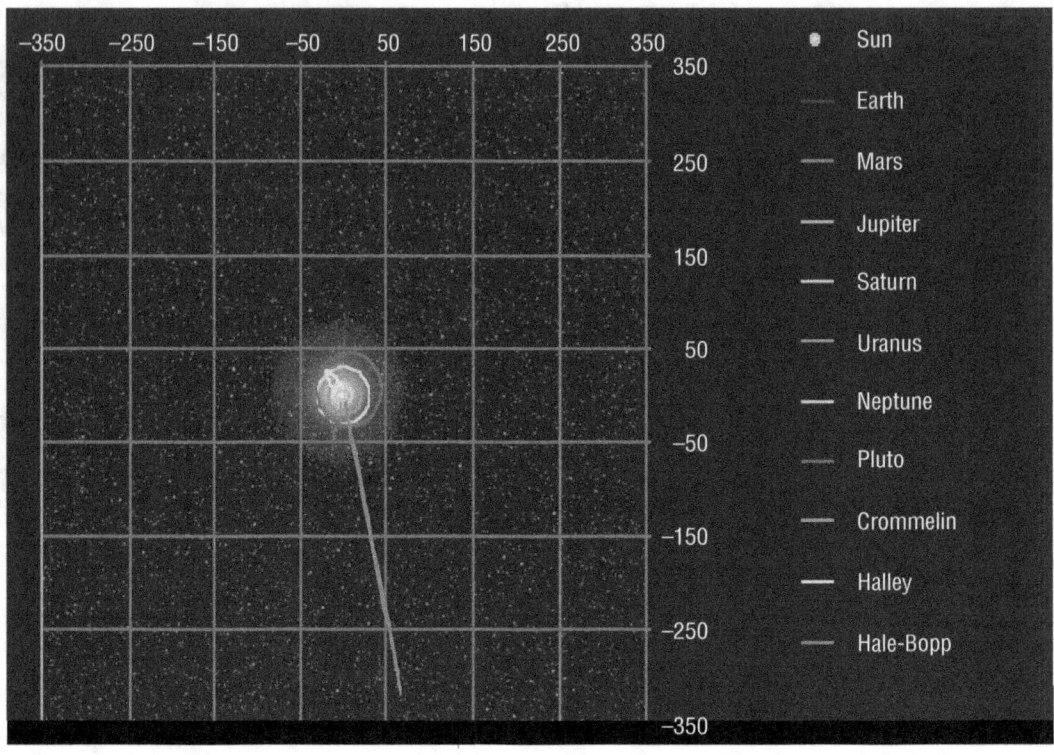

Figure 4. Orbits of representative short- and long-period comets relative to the solar system. Comet Hale-Bopp represents long-period comets.

Finally, minor planets can be organized according to their composition. A list of composition types with approximate densities is given in table 1. These densities are highly conjectural, having been deduced from samples found after atmospheric entry. Note that chondrites comprise 88 percent of the asteroid population. Previous studies indicate that ≈80 percent of impact-related deaths are likely to be caused by chondrites and short-period comets.

Table 1. Description of asteroid and comet compositions and representative predicted densities.

Asteroids	Density (g/cm^3)	Comets	Density (g/cm^3)
Chondrite	3.6	Short period	1.4
Achondrite	3.2	Long period	1.1
Iron	7.9		
Stony irons			
Mesosiderite	5.0		
Pallasite	4.3		

2.2 Earth's Impact Record

Table 2 lists the impact frequency for progressively larger near-Earth objects (NEOs). Crater diameters and terrestrial events likely to inflict comparable damage are also listed. Note that even the smallest diameter objects are capable of causing very major damage. A 23-m-diameter object can cause destruction equivalent to the nuclear weapon used at Hiroshima at the end of World War II. At the other end of the size spectrum is the Chicxulub impact, which is widely believed to have initiated an ice age at the boundary between the Cretaceous and Tertiary periods and the consequent extinction of over 50 percent of the then existing species of flora and fauna, including the dinosaurs.

Table 2. Comparable terrestrial events for NEOs of various diameters.[2]

NEO Diameter	Yield (TNT Equivalent)	Impact Frequency (Per Myr)	Crater Diameter	Comparable Terrestrial Event
2 m	500 ton	250,000	35 m	Minimum damaging earthquake (M=5)
4 m	4,500 ton	69,000	75 m	Largest chemical explosion (Heligoland Fortifications, 1947)
6 m	20,000 ton	28,000	120 m	Atomic bomb explosion (Hiroshima, Japan, 1945)
23 m	1 Mton	2,700	450 m	Typical hydrogen bomb explosion (1 Mton)
55 m	11 Mton	540	1.1 km	Barringer Meteor Crater, Arizona; Tunguska explosion, Siberia, Russia
250 m	1,4000 Mton	35	5 km	Gardnos, Norway; Goat Paddock, Australia
500 m	10,000 Mton	10	10 km	Lake Mein, Sweden; Bosumtwi, Ghana; Oasis, Libya
1 km	87,000 Mton	2.9	20 km	Haughton Dome, Canada; Rochechouart, France; Ries Crater, Germany
1.5 km	310,000 Mton	1.4	31 km	Total annual energy released from Earth (seismic, volcanic, etc.)
10 km	8.7E7 Mton	0.007	200 km	Sudbury, Canada; Vredefort, South Africa; Chicxulub, Mexico

Figure 5 illustrates the location of 145 known impact craters distributed around the world. The actual number of Earth impacts is thought to be much higher, but most of the evidence has been destroyed or covered by geological processes and vegetation. Note that the majority of the craters are <50 Ma (Mega annum, or million years) old. Additionally, most crater diameters are in the 50- to 100-km range. This evidence supports the theory that wind and water erosion, seismic events, vegetation, and the like are constantly erasing crater sites. Somewhat perversely, craters >100 km are not always simple to find because their effects are so widespread as to not be easily recognized as impact craters. For instance, the Chicxulub crater was eventually only identified from radar density mapping performed by a petroleum company owned by the Mexican government. Sometimes there is circumstantial evidence that indicates that a major impact crater is present. In the case of Chicxulub, a large number of sinkholes were found around the periphery of the impact crater. As an interesting side note, these sinkholes contained potable water, without which it may have been impossible for the Spaniards to explore that portion of the continent in the 1700's.

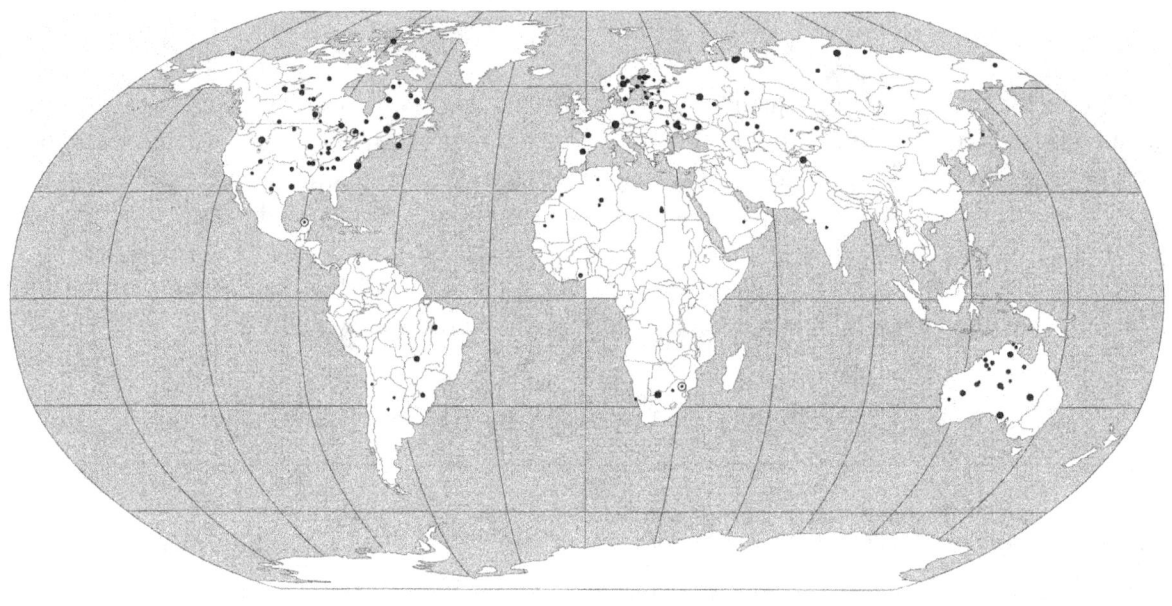

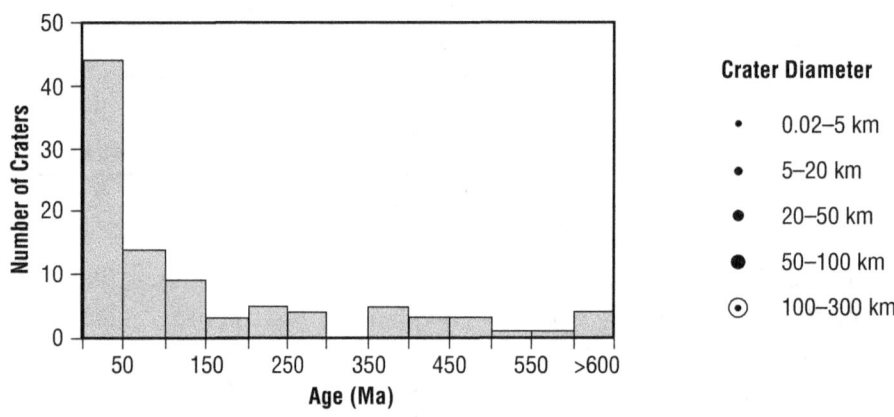

Figure 5. Location of known impact craters noting diameters as of 1998.
Age distribution of these craters is also included.[2]

Figure 5 also illustrates which parts of the Earth have been subjected to the most thorough search for evidence of extraterrestrial impacts. Many more impacts have been found in Europe, North America, and Australia than in other regions. There is no obvious reason why these continents would have received a higher impact flux than the others. It seems likely that impact structures exist in equivalent numbers, but as yet, undiscovered, on the other continents. Finally, note that few impact structures have been found underwater. It is expected that cratering is mitigated by the cushioning effect of the oceans and that the erosion rate is higher for submerged craters. In addition, it is clearly more difficult to find craters in deep water.

The last significant impact on Earth was the Tunguska event of 1908. This impact is believed to have been caused by a 30- to 60-m object that detonated at a height of ≈8 km above the Earth's surface. The estimated blast point is illustrated in figure 6. Investigators who explored the area during a series of expeditions between 1958 and 1965 carefully recorded the direction in which trees had fallen as a result of the blast. These directions are mapped in figure 6; they clearly indicate the location of the center of the event.

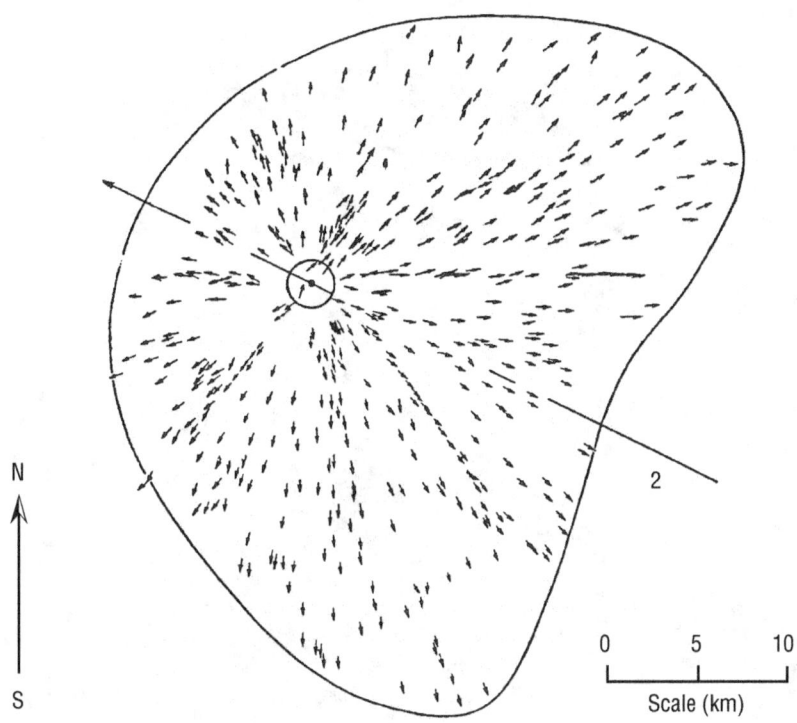

Figure 6. Projected area affected from the Tunguska blast of 1908. Arrows depict the location and direction trees were knocked down from the blast.[3]

Most strikes by large NEOs do not reach the Earth's surface. Instead, the combination of heat and stress, which the object experiences as it travels at very high speeds through the atmosphere, usually causes it to disintegrate explosively. Unfortunately, modeling and empirical evidence suggest that the heights at which such explosive blasts are most likely to occur are similar to those determined—by nuclear weapons experts—to cause maximum surface damage.

Figure 6 indicates the total ground area affected by even this relatively small object. Living creatures inside this area are not thought to have survived the event. Fortunately, Tunguska is an unpopulated area in Russian Siberia. A similar strike in a populated area would have caused widespread devastation. Consider figure 7—the Tunguska event superimposed over Madison County, Alabama, in the United States of America (USA), the authors' residence and location of MSFC. An impact of this magnitude would devastate the county, killing the majority of its 250,000 inhabitants. Superimposed over a more densely populated area, such as a large city like New York City or London, the devastation would cause the deaths of several million people.

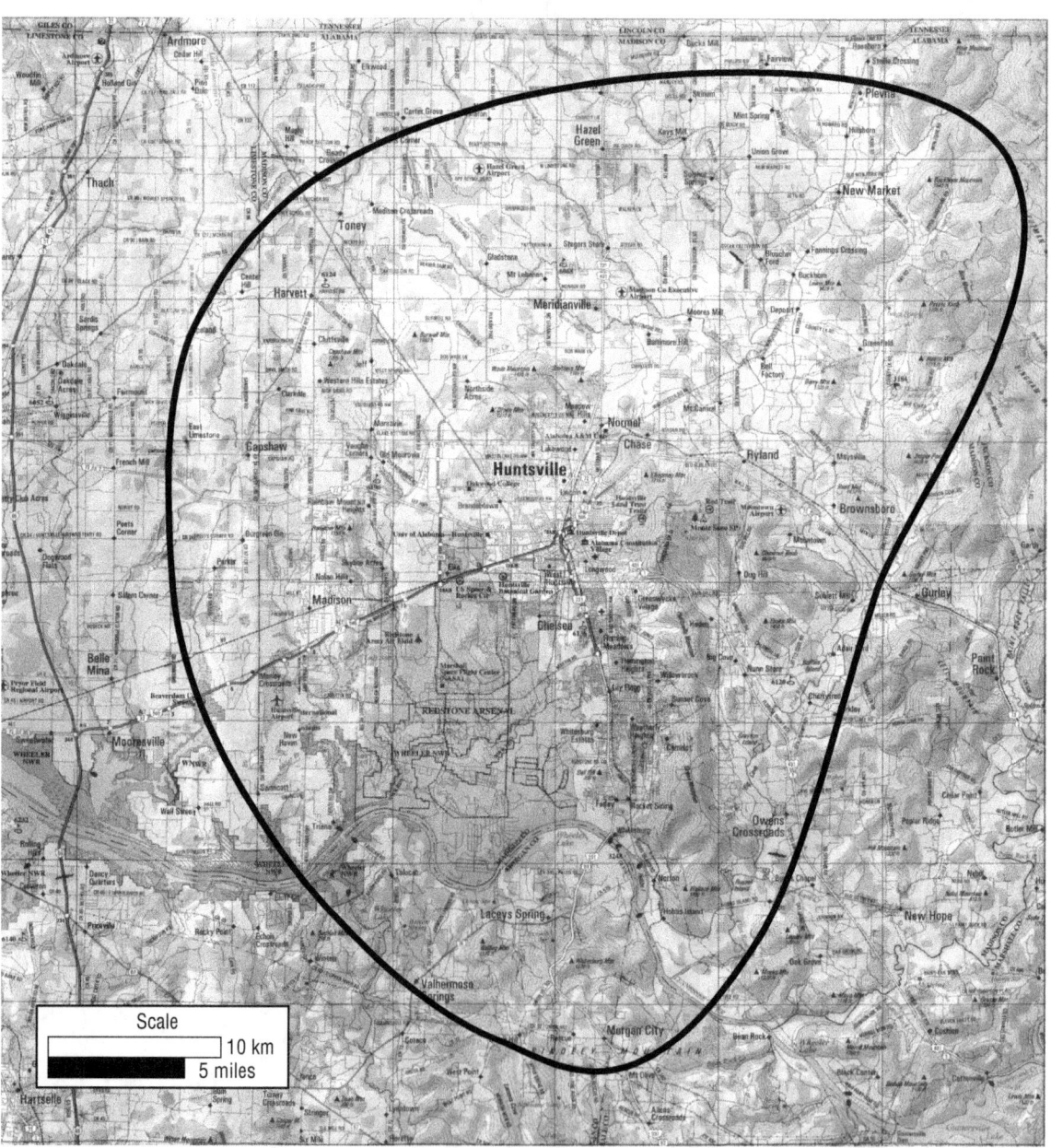

Figure 7. Tunguska impact area superimposed over Madison County, Alabama, USA. Several hundred thousand casualties can be expected from such an impact.

Figure 8 depicts an aerial view of the Barringer meteorite crater in Arizona, USA. It is ≈1 mi in diameter and ≈570 ft deep and is believed to have been caused by an object ≈150 m in diameter that impacted between 25,000 and 50,000 yr ago. This object did survive the transit through the atmosphere and physically impacted the Earth's surface. This scenario is characteristic of large nickel (Ni)-iron (Fe) asteroids that can survive the thermal and stress experienced during atmospheric entry.

Figure 8. Aerial view of the Barringer Impact Crater in Arizona, USA.[4]

The projected devastation from this impact is shown in figure 9. As can be seen, human injuries and fatalities are expected up to 24 km from the impact point. Additionally, the impact would cause hurricane-force winds with resulting damage up to 40 km away. Although this damage is less than that expected from the airburst of a non-Ni-Fe object of similar diameter, the destruction is still far from trivial.

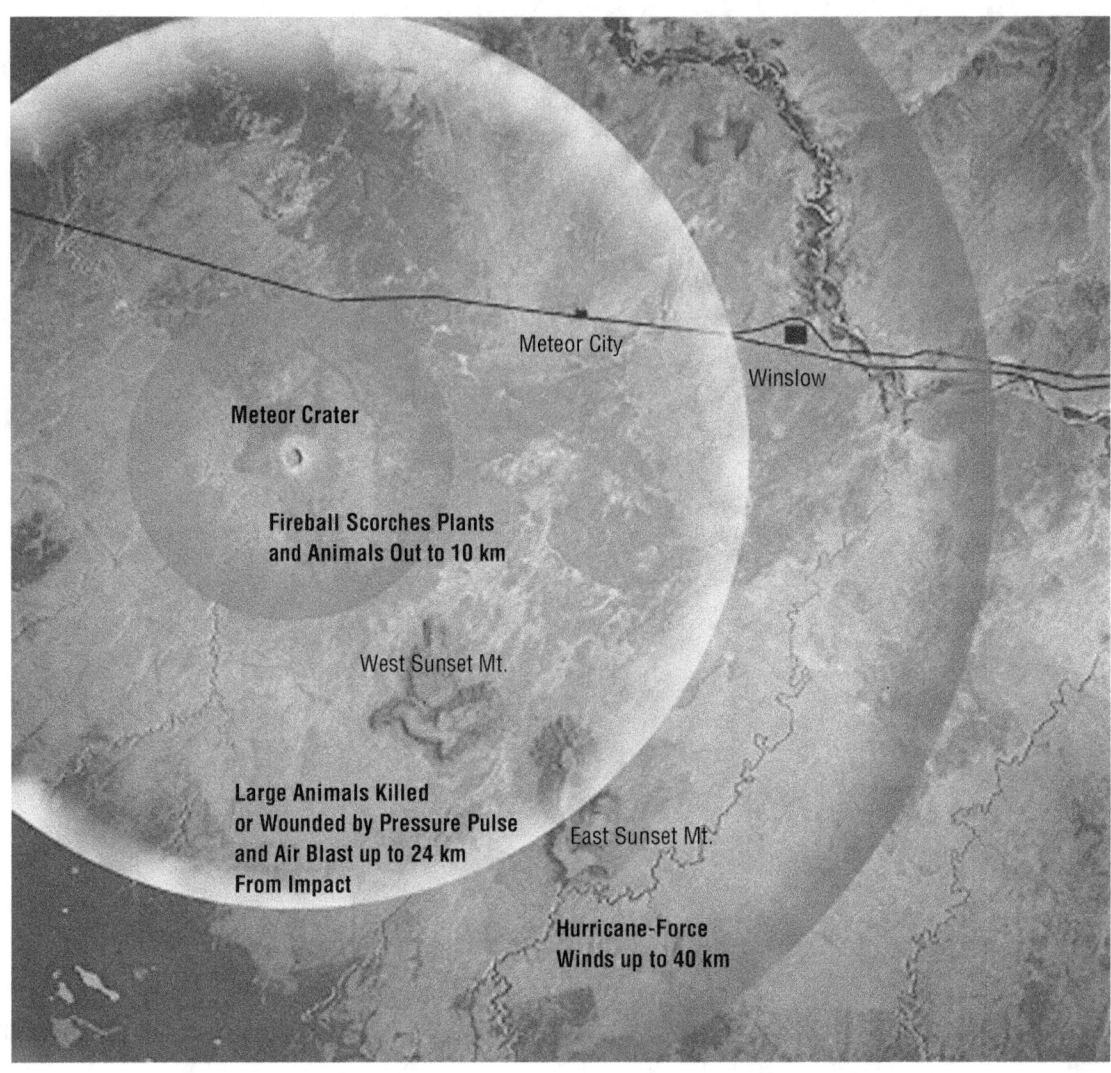

Figure 9. Calculated impact areas from the Barringer meteor.[4]

Since impacts are concealed or erased due to natural processes, consideration must be given to other types of evidence in order to determine the Earth's impact history. Figure 10 shows the estimated percentage of all species that were driven to extinction as a function of time. Many of the spikes in this graph coincide closely in time with major known impact craters. Other peaks in this graph coincide with possible stratigraphic evidence of impact—material in the geologic strata that could be due to ejecta distributed around the world following an impact.

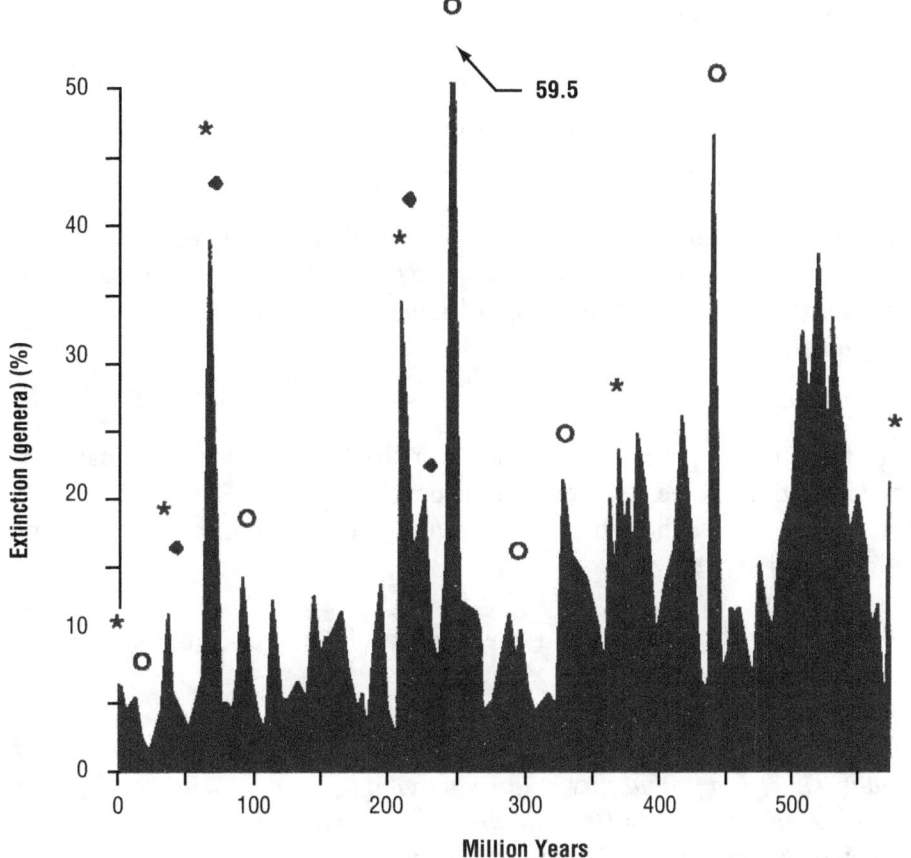

Figure 10. Extinction events versus time according to the fossil record.[5]

Possible additional evidence of impacts is provided by the record of occasional geomagnetic field reversals contained in frozen lava flows in the Earth's crust. The Earth's magnetic field periodically reverses naturally; however, there is evidence of several reversals that cannot be explained by the normal sequence of periodic changes. The Earth's field could be reversed by a sharp impact of sufficient magnitude; the mechanism is similar to that which acts when a ferromagnet is struck with a hammer to realign its magnetic field.

2.3 History Related to Near-Earth Objects

The concept that the Earth is threatened by impacts from space is not universally accepted. It is entertaining to look at some remarks made through the ages concerning impacts from extraterrestrial objects. The first quote is frequently attributed to President Thomas Jefferson:

"I could more easily believe two Yankee Professors would lie than that stones would fall from heaven."

—President Thomas Jefferson, 1807

To be fair, it is unclear whether President Jefferson actually said these words. Additional comments by him in the same time period suggest that he was more open to the audacious theory that meteoroids were of extraterrestrial origin.

Carl Sagan considered the idea of moving asteroids and comets—both for resource utilization and planetary defense—as a potential unifying endeavor for humanity:

"Since hazards from asteroids and comets must apply to inhabited planets all over the Galaxy, if there are such, intelligent beings everywhere will have to unify their home worlds politically, leave their planets, and move small nearby worlds around. Their eventual choice, as ours, is spaceflight or extinction."

—Carl Sagan, Pale Blue Dot, 1994

Here Sagan suggests that planetary defense from NEOs is a strong justification for the continued exploration of space. Additionally, he believed that the development of systems needed to defeat this threat should not be shouldered by one nation, but would require participation of all of humanity. This premise will be addressed in section 9.

Edward Teller, renowned physicist and developer of the hydrogen bomb, also commented on the threat posed by asteroids and comets:

"Here is the situation that, to my mind, is a scandal, and I think people can understand that it is a scandal: There is a probability of a few percent in the next century of the arrival of a stony asteroid ... approximately 100 m in diameter. It is a practical certainty that ... it will come completely unannounced. We won't have any indication of it. Yet such an object is apt ... to do a lot of damage. ... Just in dollars it could be billions, and in lives it might reach millions."

—Edward Teller, LLNL, 1995

Teller illustrates the threat represented by NEOs. Some astronomers have strongly suggested that the threat necessitates a larger investment than is currently being made in the business of searching for and categorizing NEOs. Teller indicates that under current circumstances there is likely to be little or no warning before a catastrophe occurs.

An effective search program clearly requires the use of telescopes in both the Northern and Southern Hemispheres. Efforts in the Southern Hemisphere were adversely affected by the Australian government's decision to withdraw funding. At the time, the Australian Minister for Science made the following comment:

"I'm not going to be spooked or panicked into spending scarce research dollars on a fruitless attempt to predict the next asteroid. I'm just not convinced that the hype and alarm and even fear-mongering is enough to justify an instant investment."
—Peter McGauran, Australian Minister for Science,
60 Minutes Interview, 2002

This action threatened to end all survey efforts in the only participating Southern Hemisphere country. Today, the Australian survey efforts continue but are funded by sources in the United States and Europe. One prominent member of the Australian survey team responded to the Minister's actions in a direct manner.

"The dinosaurs did not have a space program. That's why they died."
—Duncan Steel, 2002 former member, Anglo-Australian
Near-Earth Asteroid Survey (AANEAS),
now Professor, University of Salford, England

The continued efforts to survey the population of NEOs are summarized in appendix A. In considering popular skepticism about the threat of Earth impacts, it is instructive to consider the development of human knowledge about comets and asteroids. Table 3 contains a list of notable and relevant scientific discoveries. As can be seen, our understanding of asteroids only dates back about two centuries. Although knowledge of the existence of comets predates written records, most of human history comets were revered (or feared) as omens and not regarded as objects of scientific curiosity. The Alvarez theory, identifying the Chicxulub impact as being responsible for the Cretaceous-Tertiary (K-T) extinction, has only been widely publicized since 1980.

Table 3. Timeline of scientific discoveries relevant to humanity's knowledge of asteroids and comets.

300 B.C.	Aristarchus theorizes circular orbits around Sun
1543	Copernicus proposes Sun-centered system
1608	Hans Lippershey invents the telescope
1609	Galileo makes first astronomical telescope observations
1609	Kepler develops first two laws of planetary motion
1687	Newton publishes *The Principia*
1705	Halley reports findings on cometary trajectories
18th C	Existence of asteroid belt theorized by Bode, Kant, et al.
1794	Chladni suggests extraterrestrial origin of meteorites
1801	Giuseppe Piazzi discovers Ceres
1932	Reinmuth discovers 1862 Apollo
19th/20th C	Discovery of Atens, Apollos, etc.
1980	Alvarez theorized asteroid impact for K-T extinction
1994	Recorded impact of Comet Shoemaker–Levy 9 into Jupiter

While an impact between a significantly sized NEO and Earth has not been directly recorded, a cometary impact with Jupiter was recently observed. On July 16, 1994, Comet Shoemaker–Levy 9 was observed impacting Jupiter by the approaching Galileo spacecraft. Results of the impact were recorded by telescopes all over the world and in Earth orbit. David Levy, co-discoverer of the comet, documented this historic event in reference 6.

2.4 Measuring the Near-Earth Object Population

To determine the overall threat posed by the solar system's asteroids and comets, one must develop a proper understanding of the populations of these two types of bodies. Unfortunately, neither population is well understood. It is believed that orbital parameters are currently known for only ≈10 percent of the total NEO population. Also, since larger objects are easier to detect, our knowledge of the known NEO population is biased toward these larger objects.

Using the known NEO population, the relative location of these objects can be plotted together with the orbits of the inner planets. This plot is shown in figure 11 and gives the location of all known objects on March 2, 2002. The green circles are minor objects that are not considered candidates for Earth impact. The red circles are minor objects that have perihelia <1.3 au. Blue squares represent periodic comets. The planets are shown as crosshair circles on their orbits. Figure 11 illustrates the large population of NEOs around the Earth at any time, however, does create a misleading impression. Due to the finite pixel size, the inner solar system—particularly the asteroid belt—appears to be full of NEOs; in fact, of course, it is overwhelmingly empty.

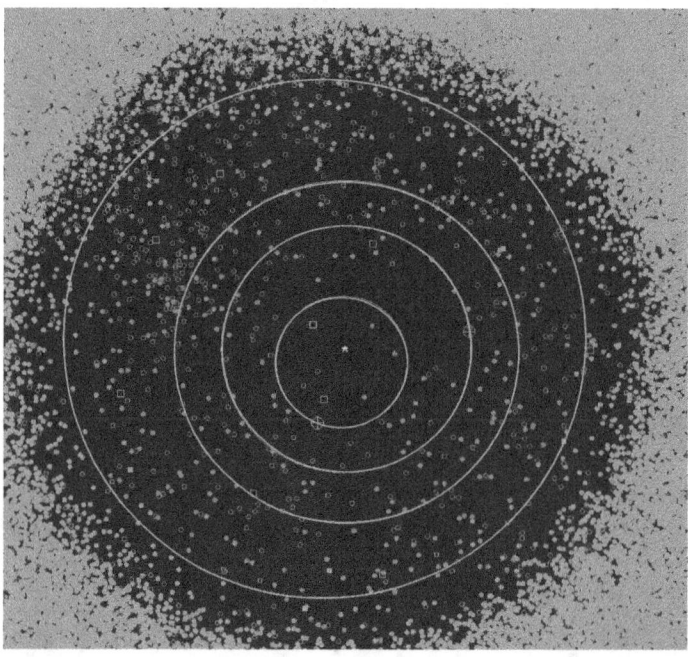

Figure 11. Location of known minor planets on March 2, 2002, plotted relative to the inner planets. NEOs are red, other asteroids are green, and comets are blue.

Our knowledge of the NEO population has increased significantly in the past few years. The number of known near-Earth asteroids (NEAs) is plotted shown as a function time in figure 12. The rapid increase in the number of objects identified over recent years can be explained in a number of ways:

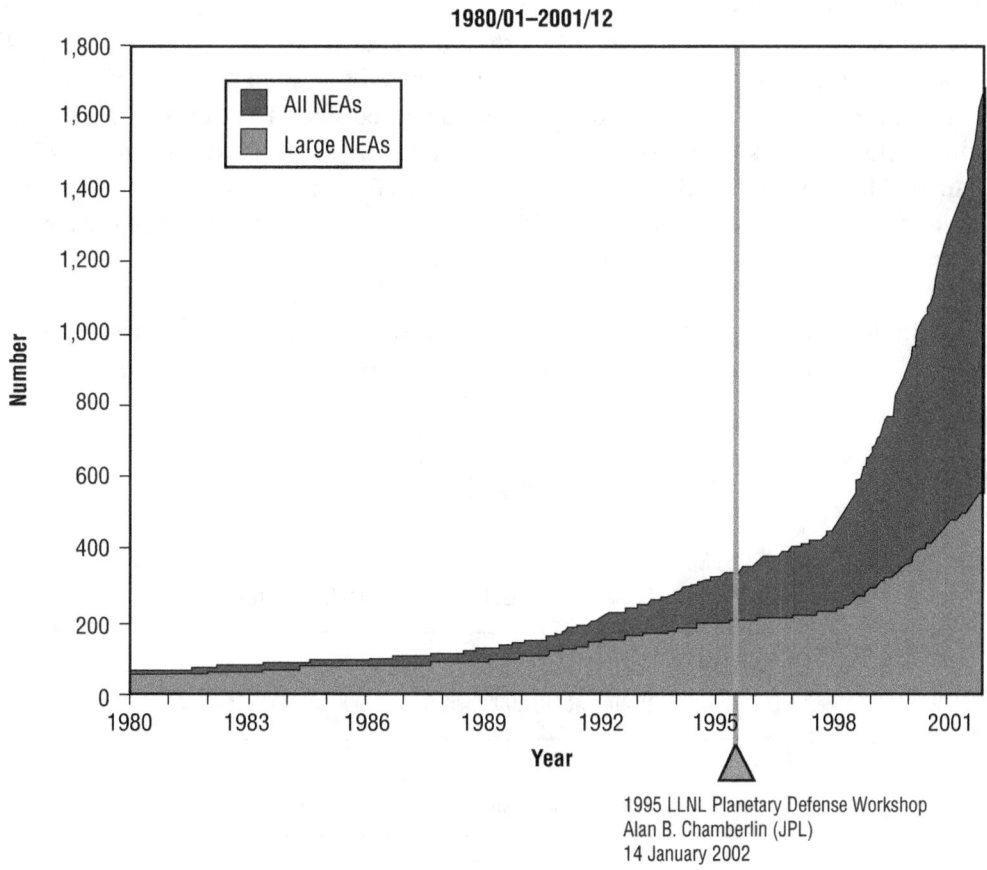

Figure 12. Number of known NEAs versus time. Note the rapid increase in discoveries in recent years due to the use of CCDs and increased interest in the asteroid and comet threat.[7]

(1) The recent acceptance of the Alvarez hypothesis, and the theory that other Earth impacts have also affected the planet in the past, have led to a significant increase in the time and resources available to locate these objects.

(2) The development of charged-coupled devices (CCDs) has computerized the previous manual process of searching for new celestial objects. Before the advent of CCD technology, such searches were conducted by the human study of photographic plates.

(3) Use of the Internet has facilitated international coordination and data sharing from sky surveys.

Note that, over recent years, the total number of asteroids located has increased more rapidly than the number of large asteroids. This is due to the fact that CCDs enable the detection of smaller objects more than was previously thought possible.

Several movies and television programs have been released in recent years, giving the general public an appreciation, however scientifically distorted, of the NEO threat. The increase in NEO detection rates, coupled with the increase in public awareness, has raised the level of coverage given in the general media to this threat. Recently, major news outlets have published several articles detailing the Earth's close encounters with NEOs; table 4 gives some details. To our knowledge, several asteroids have passed the Earth at a distance of less than twice the Moon's orbital radius from the Earth in the past year. In at least one instance, the detection was made after the asteroid had already passed the point of closest approach.

Table 4. Recent near misses by comets and asteroids. By comparison, the distance between the Earth and the Moon is ≈240,000 mi.

2001 YB5	300- to 400-m diameter January 7, 2002, flew 375,000 mi from Earth
2002 EM7	70-m diameter March 8, 2002, flew 288,000 mi from Earth
2002 NY40	800-m diameter August 18, 2002, will fly 330,000 mi from Earth
2002 NT7	1.2-km diameter, 28 km/s Will fly by Earth on February 1, 2019

2.5 Damage Mechanisms

The potential damage mechanisms resulting from an impact are several and varied; some are straightforward while others affect the Earth's ecosphere in a more indirect manner. Some of the mechanisms presented here are a little speculative, but this is to be expected. There is, fortunately, not a lot of applicable empirical data upon which to draw.

The mechanisms can be broadly categorized as either directly or indirectly linked to the impact itself. Within the directly linked category are the tsunami and the blast wave. Within the indirectly linked category are the effects of releasing large quantities of dust, nitric oxide, and water vapor into the atmosphere, as well as the possible geopolitical outcome of an impact.

Before considering the damage mechanisms, it is important to discuss the two types of impact events that can occur—air burst and surface impact. During passage through the atmosphere, an incoming body will experience a considerable ram pressure force on its leading surface. This force, in addition to decelerating the body, will place it under considerable internal stress, with some resulting strain. A sufficiently friable body, such as one comprised largely of stony material, is likely to fragment. The potential energy imparted by the stress forces being suddenly released causes the fragments to fly apart explosively. This large-scale disintegration, while the body is still at some altitude, is referred to as an airburst. A body with a more solid internal structure, particularly one with a largely metallic composition, will be able to withstand the ram pressure-induced stresses and will strike the surface—either land or water—largely intact. This is referred to as a surface impact.

Although the kinetic energy released locally is clearly greater for a surface impact, damage can extend over a wider area from an airburst, particularly one at the optimum altitude. The Tunguska meteorite produced an airburst at an altitude of ≈8 km while traveling at ≈20 km/s. Although the level of destruction immediately below the airburst was less than would have been produced by a surface impact; i.e., no crater was produced, the cumulative damage over an extended area was greater.[3]

Now, regarding the resulting damage mechanisms, both airbursts and surface impacts can give rise to all of those detailed in sections 2.5.1 through 2.5.6. There will undoubtedly be some difference in the magnitude and extent of a particular mechanism if produced by an airburst rather than by a surface impact, but at the present high level of detail, such differences are not important.

2.5.1 Tsunami

Both an oceanic impact and an airburst over the ocean will generate high-energy water waves. These waves are two-dimensional disturbances whose height diminishes inversely with distance from their point of origin. The key to development of a tsunami is the relationships between wave speed and water depth, and among wave energy, height, and speed. Shallower water results in a lower speed, but slower waves become steeper due to simple energy conservation. As a wave approaches the continental shelf, the shallower water slows it and increases its height. By these mechanisms, a wave of moderate amplitude in the deep ocean can increase in height by an order of magnitude as it comes ashore. With increased wave height, the potential for causing damage ashore increases proportionately.

Tsunamis have caused great damage and loss of life in coastal areas, particularly low-lying regions. In July of 1998, a 30-ft-high tsunami came ashore in Papua, New Guinea, killing more than 2,100 people. The cause of the tsunami was determined to be an underwater landslide that took place more than 2,000 mi distant.

Recent modeling studies indicate that the surface impact of a 400-m-diameter body—at any point in the Atlantic Ocean—would devastate both the American and European/African coastlines with final wave heights in excess of 60 m.[3] The height of deep-water waves 1,000 km distant from the impact point of a soft stone meteor of varying size[8] is shown in figure 13. Figure 14 shows the height of a deep-water wave 1,000 km distant from the impact point of an Fe meteor of varying size.[8]

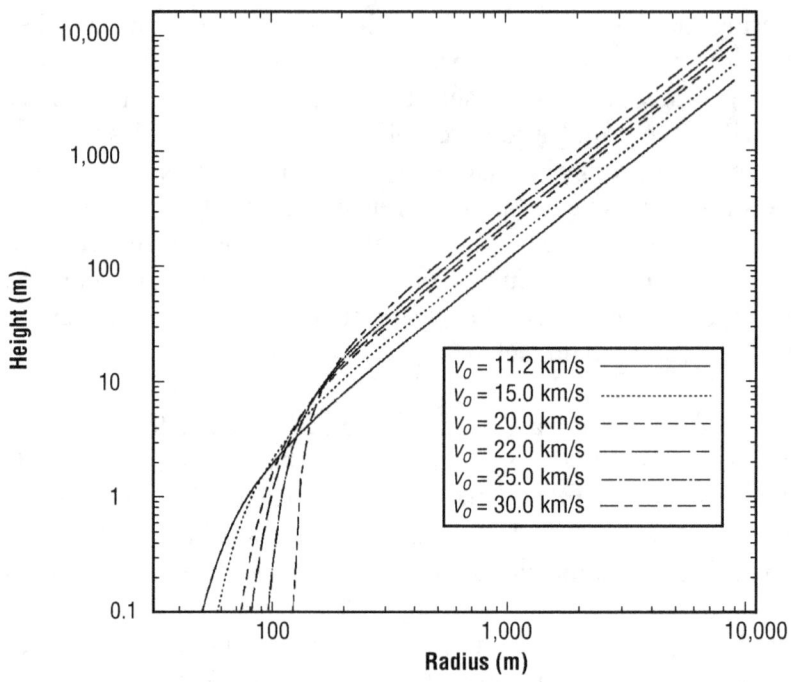

Figure 13. Deep-water wave height at 1,000 km distance versus initial meteor radius for soft stone meteor.

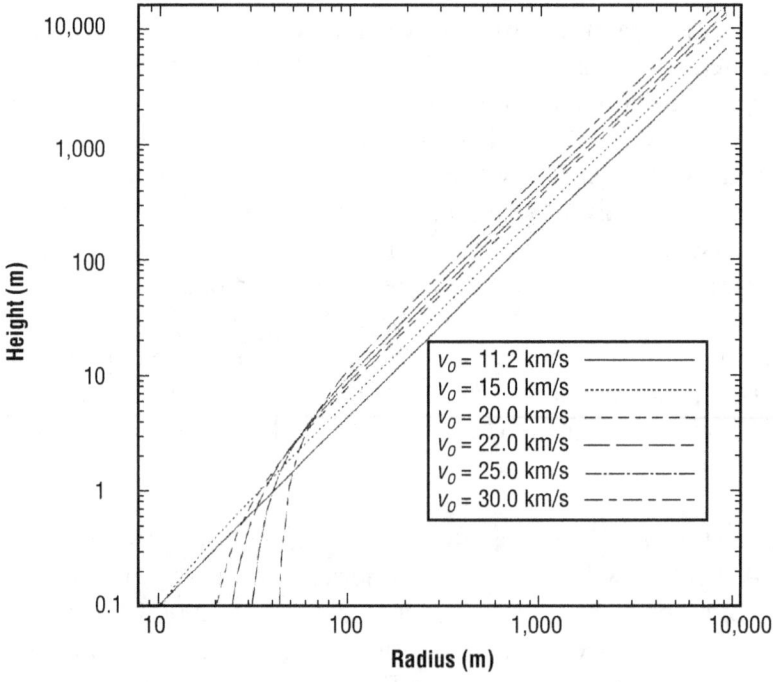

Figure 14. Deep-water wave height at 1,000 km distance versus initial meteor radius for Fe meteor.

2.5.2 Blast Wave

Blast waves are produced by both airbursts and surface impacts. The destructive potential of the wave is determined by the total explosive energy of either the airburst or the surface impact as well as the altitude at which the explosion occurs.

The blast wave consists of a shock wave followed by a substantial wind. The shock is characterized by the peak overpressure. Even a 2-psi (13.972-kPa) overpressure will create a severe hazard due to flying debris. A 4-psi (27.944-kPa) overpressure corresponds to hurricane-force winds of 70 mph.

Figure 15 (taken from ref. 9) consists of a graph showing the total damage area, defined as the area that experiences an overpressure of 4 psi or higher, as a function of the impact energy. Within the graph, the following abbreviations are used:

SP   = short-period comet
LP   = long-period comet
Carb = carbonaceous chondrite.

The surface explosion curve is derived on the basis that only 3 percent of the impact energy goes into shock waves; i.e., $\varepsilon = 0.03$.

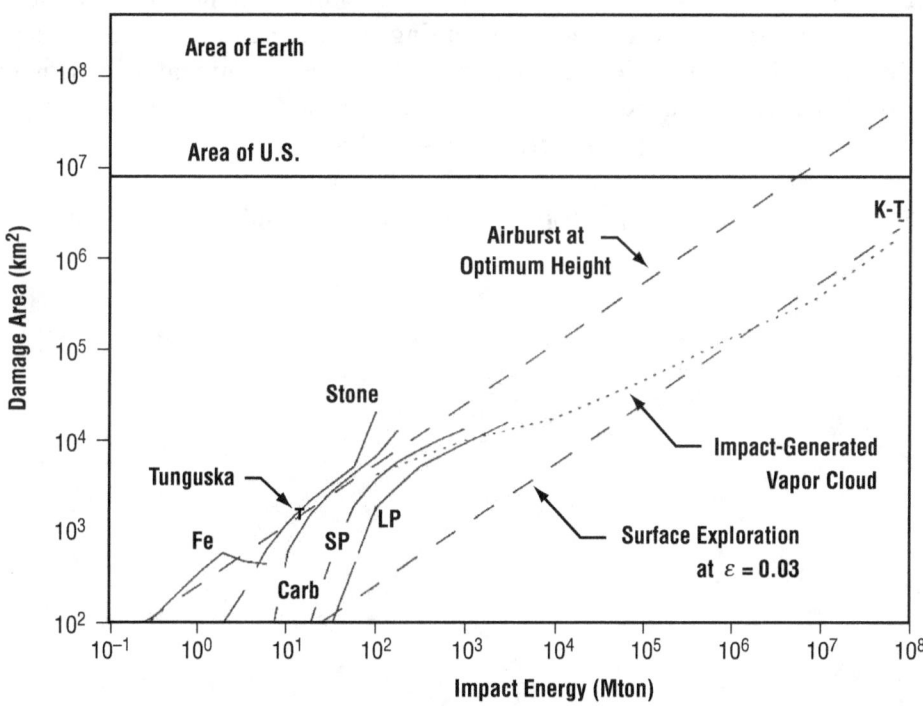

Figure 15. Blast wave damage versus impact energy.

2.5.3 Atmospheric Dust Loading

Submicron dust reaching the stratosphere as a result of an impact will adversely affect the regional and global climate as well as disrupting important biochemical mechanisms, such as photosynthesis.

The sequence of events is complex and four distinct steps are required to model the effects of dust loading:

(1) The amount of dust lofted by the impact must be derived, based upon the impacting body's composition, its kinetic energy, and the nature of the impact.

(2) The rate at which dust disperses around the globe must be derived, taking into account the settling process—some dust will leave the atmosphere.

(3) The effect of the atmospheric dust upon Earth's radiation balance must be calculated.

(4) The effect of low light levels on surface temperature, precipitation, and photosynthesis must be established.

It is anticipated that a large impact could cause a drop of several degrees in global temperature and the loss of one or more year's crop with resulting starvation, mass migration, social disorder, and possibly warfare. Quantification is difficult without extremely complex numerical modeling. Most work to date has concentrated upon the atmospheric dust loading, resulting from the use of nuclear weapons. Modeling results indicate that a 5,000-Mton nuclear exchange could reduce continental interior temperatures to –25 °C within as little as 2 wk. In practice, heat transfer from ocean water would probably mitigate somewhat, resulting in minimum temperatures of only about –5 °C.[5]

Figure 16 shows the density of atmospheric dust and the resulting optical depth as functions of impact energy.[9]

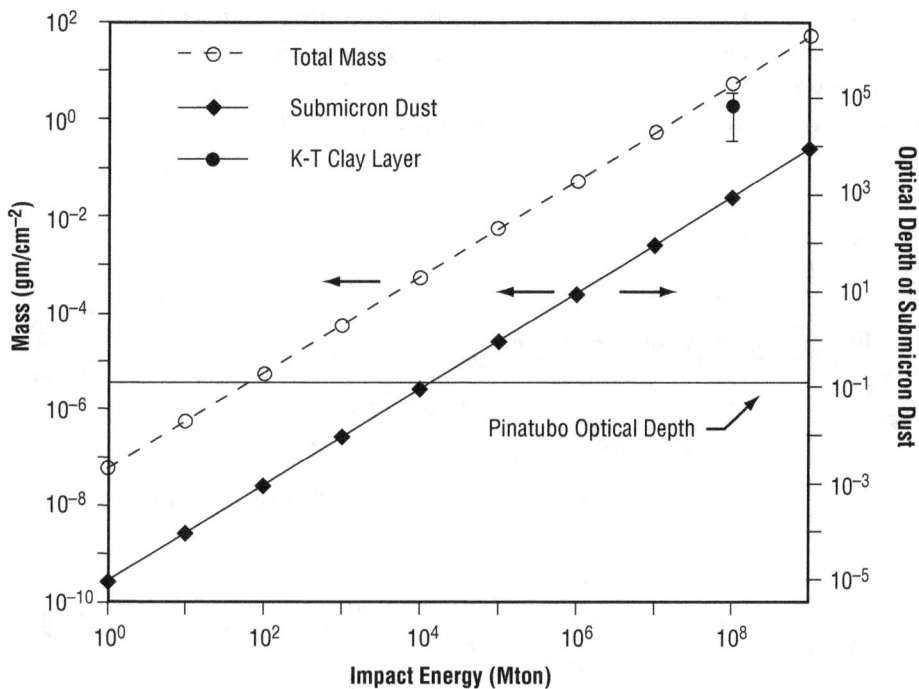

Figure 16. Density and optical depth of atmospheric dust versus impact energy.

2.5.4 Nitrogen Oxide Production

Four significant mechanisms for the production of nitrogen oxides (NOs) during an impact have been identified:

(1) The formation of NO in the shock waves that accompany both atmospheric entry and explosive disintegration.

(2) If a surface crater is created, NO forms if the resulting ejecta plume moves through the atmosphere at more than 2 km/s.

(3) If the ejecta plume leaves the atmosphere, further NO will be formed within the subsequent reentry shocks.

(4) In the case of an impact from an extremely large body, the hot ejecta dispersed through the atmosphere could briefly bring the local temperature up to $\approx 1,500$ K, at which point additional NO would be created directly from atmospheric gases.

The most direct consequence of elevated atmospheric NO levels are the production of acid rain. In addition to elevated atmospheric NO levels, sulfate- or carbonaceous-rich impacting bodies may also produce sulfur dioxide and carbon dioxide, respectively. Elevated NO levels in the atmosphere would also threaten the integrity of the Earth's ozone layer. An NO volume mixing ratio; i.e., volume of NO as a fraction of total air volume, as low as 2×10^{-7}, if mixed uniformly throughout the atmosphere, would render the current ozone ultraviolet screen ineffective. Figure 17, taken from Toon et al.,[9] contains a brief summary of the effects of NO on both the ozone layer and oceanic acidity levels.

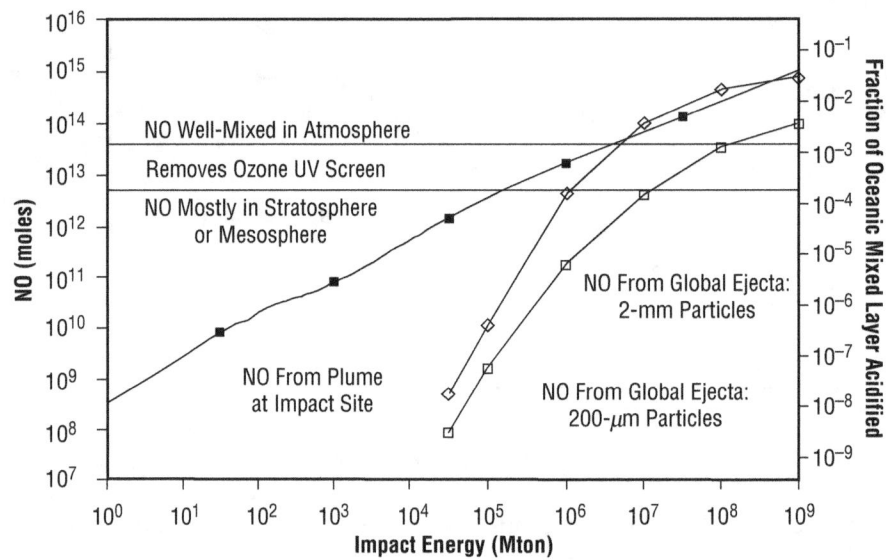

Figure 17. Blast wave damage versus impact energy.

2.5.5 Water Vapor Injection

Under normal circumstances, the upper atmosphere has very low humidity levels. A large oceanic impact would inject large amounts of water into the upper atmosphere. The resulting increase in water levels above the tropopause would give rise to a "greenhouse" effect and would substantially increase Earth's temperature. There are two mechanisms by which water can reach the upper atmosphere: direct splash to high altitude and water vaporization.

Analysis indicates $\approx$150 million kg of water would be vaporized per megaton of impact energy.[4] The current water vapor level above the tropopause is approximately 2×10^{-4} to 6×10^{-4} gm/cm^2; this means that impacts as low as 104 Mton are capable of doubling that level.

The impact of additional water vapor at high altitude is uncertain, mainly because the mechanisms by which water leaves the upper atmosphere are not properly understood. However, the resulting greenhouse effect could significantly increase global temperatures. Figure 18 shows the mass of water injected into the atmosphere as a function of impact energy.[9]

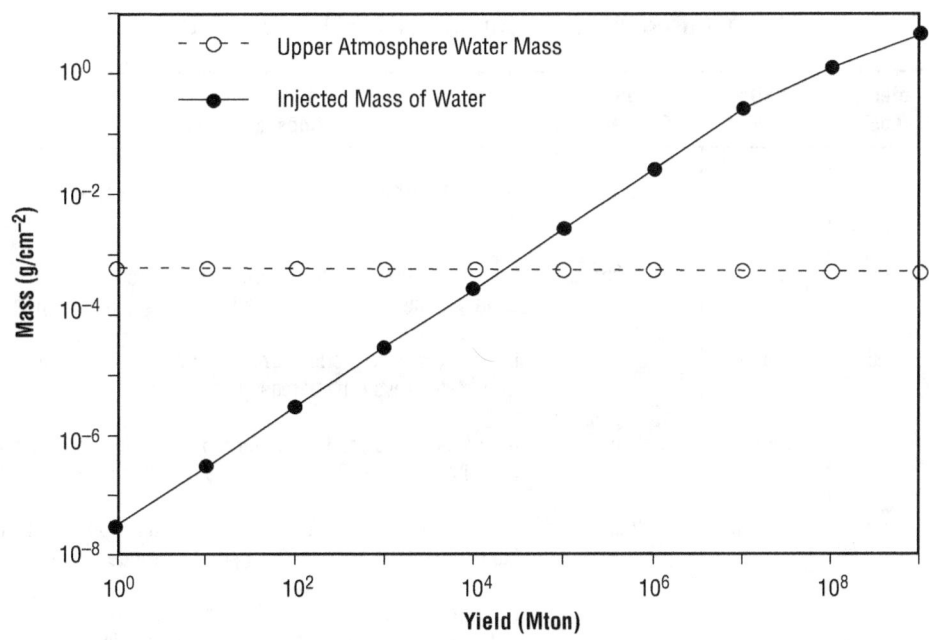

Figure 18. Mass of water lifted into the atmosphere versus impact energy.

2.5.6 Precipitate Nuclear Exchange

In recent years, several new nations have acquired the capability to produce nuclear weapons, and some of them have proceeded to develop such weapons. Although efforts are underway to prevent further proliferation of nuclear weapons technology, it is quite possible that additional nations will succeed in acquiring it over the next few years. Although possessing the weapons themselves, most of these nations do not yet have the relatively sophisticated detection systems of the better established nuclear powers. This raises the concern that an unexpected impact might be mistaken for a nuclear strike. If this scenario were to occur in a region of high tension, it could trigger a retaliatory nuclear strike.

Although nations such as the United States would probably be able to rapidly distinguish an impact event from a nuclear weapon detonation, it is entirely possible that this information would attract little credence in a region of high international tension. With its public demanding retaliatory action against the supposed perpetrator, a national government might discover that events had acquired a grim and irresistible momentum.

This recent development adds urgency to the need for more thorough identification of threatening bodies, more accurate tracking of them, and the development of effective protection techniques.

2.6 The Credibility Problem

Considering the research described above, one might ask why the danger posed by asteroids and comets is not given more public attention. The answer to this question is somewhat complex. First, it is important to distinguish between impacts from small and large objects. Examination of table 5 illustrates

Table 5. Consequences of impact by NEOs of various sizes.[10]

Yield (Mt)	Interval LogT	NEO Diameter	Crater Diameter	Consequences
<10	–	–	–	Upper atmosphere detonation of stones and comets; only irons (<3%) reach surface.
$10–10^2$	3.0	75 m	1.5 km	Irons make craters; stones produce airbursts. Land impacts destroy city-sized areas; e.g., Washington, DC.
$10^2–10^3$	3.6	160 m	3 km	Irons and stones produce groundbursts; comets produce airbursts. Impact destroys urban areas; e.g., New York City.
$10^3–10^4$	4.2	350 m	6 km	Impacts on land produce craters; ocean tsunamis become significant. Land impact destroys area the size of a small state; e.g., Delaware.
$10^4–10^5$	4.8	0.7 km	12 km	Tsunamis reach oceanic scales, exceed damage from land impacts. Land impact destroys area the size of a moderate state; e.g., Virginia.
$10^5–10^6$	5.4	1.7 km	30 km	Land impact raises enough dust to affect climate, freeze crop. Ocean impacts generate hemispheric-scale tsunamis. Global ozone destruction. Land impact destroys area the size of a large state; e.g., California.
$10^6–10^7$	6.0	3 km	60 km	Both land and ocean impacts raise dust, change climate. Impact ejecta are global, triggering widespread fires. Land impact destroys area the size of a large nation; e.g., Mexico.
$10^7–10^8$	6.6	7 km	125 km	Prolonged climate effects, global configuration, probably mass extinction. Direct destruction approaches continental scale; e.g., United States.
$10^8–10^9$	7.2	16 km	250 km	Large mass extinction; e.g., K-T-type event.
$>10^{10}$	–	–	–	Threatens survival of all advanced forms of life.

the consequences of impacts by objects of various sizes. Large impacts, caused by objects with a diameter greater than the 500-m to 1-km range, produce effects that are felt across the globe. These impacts have received some attention in the popular media. Smaller objects are likely to cause devastation on a regional scale. Although more likely to occur than a larger impact, the danger from these smaller bodies has been ignored; there are several reasons for this.

First, the concept of severe destruction being caused by a collision with an NEO is very alien to most of the general public. Nobody knows of anyone who has been killed by a falling meteorite and very few people have witnessed a fall of any size. Human nature tends to naturally concentrate attention on dangers that are perceived as present in our everyday lives; i.e., automobile accidents, fires, etc., and excludes those that are more exotic. With all the threats facing humanity in the early 21st century, the subconscious decision to avoid facing such a nonobvious threat could even be seen as a defense mechanism against being psychologically overwhelmed.

As mentioned previously, our knowledge of comets and asteroids is relatively recent. The idea of asteroids and comets colliding with the Earth and causing widespread devastation has had even less time to take root. Also, although careful searches have recently yielded indirect evidence, direct evidence of impacts from celestial objects is not readily found in the historical record. However, several historical incidents could be interpreted as asteroid impacts. Sources as diverse as the Christian Bible and Maori tribal records from New Zealand make reference to catastrophic events that could have been caused by large impacts.

In many respects, collisions with asteroids and comets are a bigger threat now than in any other time in history. This can be argued by considering human population growth over the past century, as depicted in figure 19.

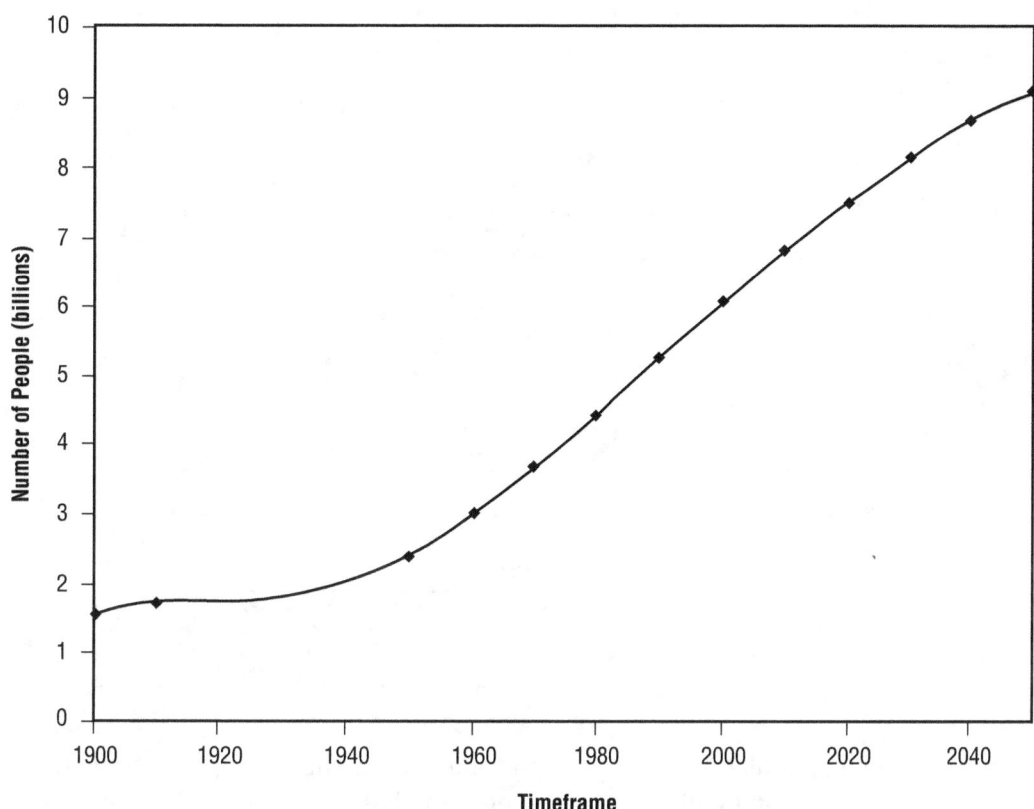

Figure 19. Predicted world population in the last century. Population is extrapolated through the middle of this century.[11]

At the time of the last collision with a substantial object (Tunguska), the Earth's human population numbered between 1 and 2 billion. Today, the Earth's population numbers above 6 billion and is rising rapidly. The increased population density means that there are far fewer remote places where an impact event, like that at Tunguska, could occur without causing significant loss of life. As mentioned in the section on damage mechanisms, even a small impact can cause a large tsunami that would affect much of the Earth's shorelines. In fact, consideration of figure 20 shows that a majority of the Earth's population actually lives near these shorelines.

An impacting asteroid would probably create large numbers of charged particles during its travel through the atmosphere. These particles would probably have a severely detrimental effect on the global electronic and communications infrastructure. Atmospheric dust loading would decrease light reaching Earth's surface, placing viable arable land at a very high premium. Our complex and interdependent technological society, usually well equipped to deal with an isolated crisis, would probably be very vulnerable to such a varied and large number of coinciding problems. Even impacts from relatively small objects—a few tens of meters in diameter—pose a much higher threat than any other time in history.

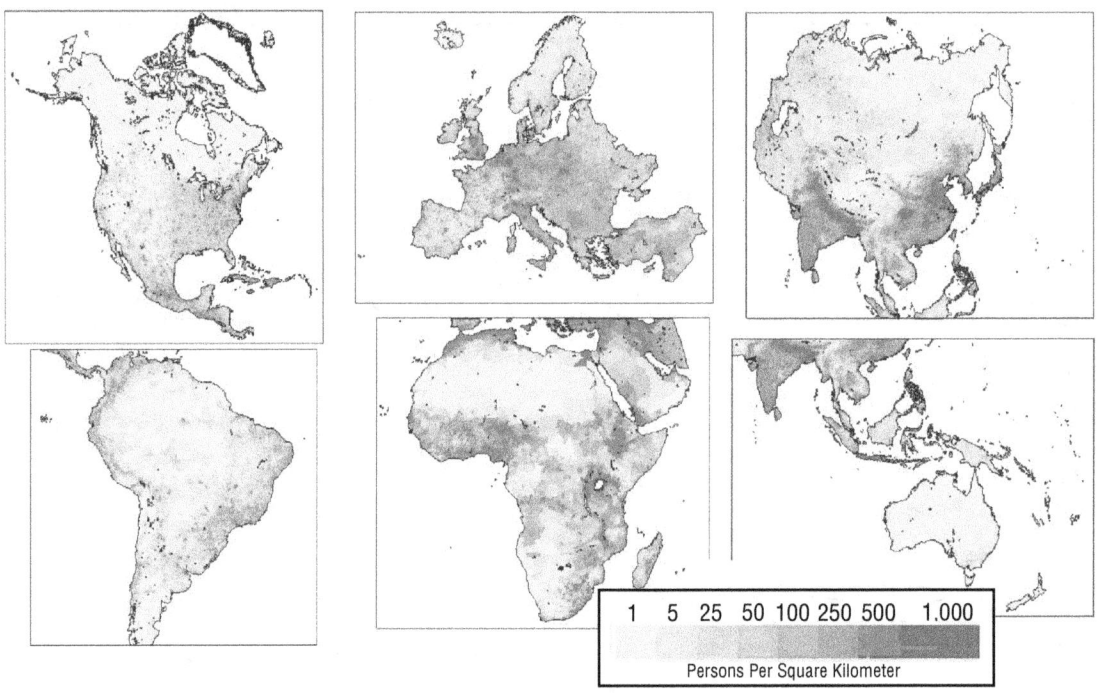

Figure 20. Human population density graph for all continents (except Antarctica). Greenland and Iceland are not represented as well as some Pacific islands.[11]

2.7 The Torino Scale

During an international conference on NEOs held in Turin in 1999, an attempt was made to establish a type of "Richter Scale" for categorizing the Earth-impact hazard associated with newly discovered asteroids and comets. In honor of the meeting venue, this system was named the Torino Scale. The Torino Scale utilizes numbers that range from zero to 10:

• Zero indicates that an object has a zero or negligibly small chance of collision with the Earth. Zero also categorizes any object too small to penetrate the Earth's atmosphere intact.

• Ten indicates that a collision is certain, and the impacting object is so large that it is capable of precipitating a global climatic disaster.

Categorization on the Torino Scale is based on the placement of a close approach event within a graphical representation of kinetic energy and collision probability (fig. 21).

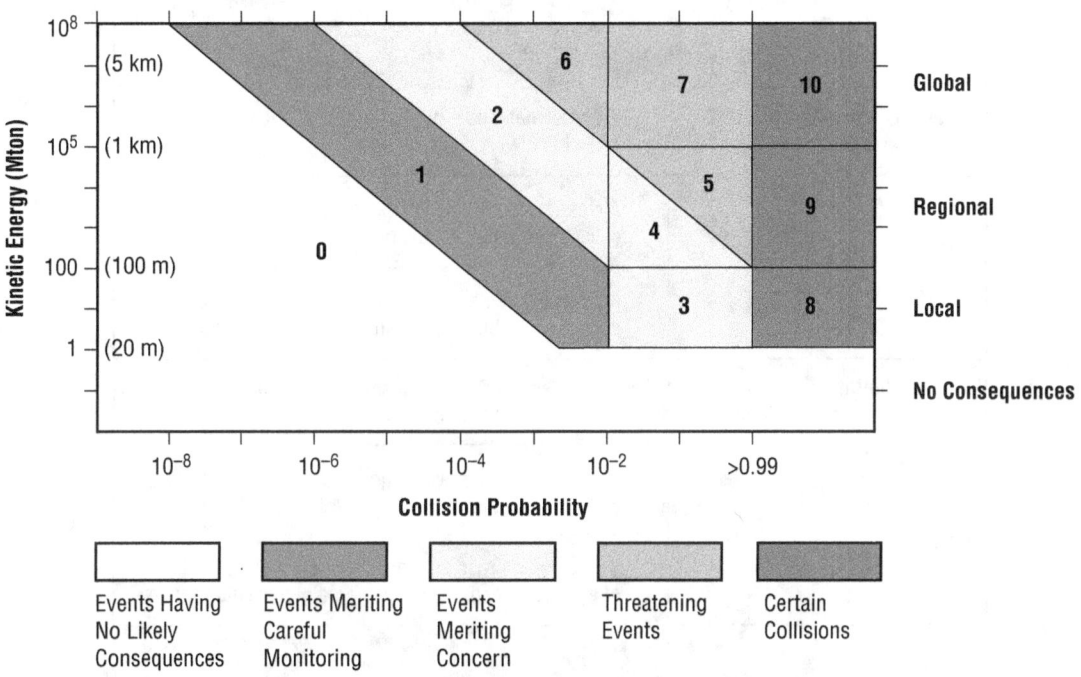

Figure 21. Illustration of the various category of threat under the Torino Scale.

Orbital predictions for newly discovered bodies are naturally uncertain. Discovery observations typically involve measurements over only a short orbital track and so, as a body's orbit characterization improves, its Torino number can change. Hopefully, it will always reduce as more information becomes available.

An object that is capable of making multiple close approaches to the Earth will have a separate Torino Scale value associated with each approach. An object may be summarized by the single highest value that it attains on the Torino Scale; no fractional or decimal values are used. The various categories within the scale are shown in figure 22. It should be noted that the Torino Scale has yet to achieve any wide-scale usage. It is mentioned here for completeness only.

TORINO SCALE Assessing Asteroid and Comet Impact Hazard Predictions in the 21st Century		
Events Having No Likely Consequences	0	The likelihood of a collision is zero, or well below the chance that a random object of the same size will strike the Earth within the next few decades. This designation also applies to any small object that, in the event of a collision, is unlikely to reach the Earth's surface intact.
Events Meriting Careful Monitoring	1	The chance of collision is extremely unlikely, about the same as a random object of the same size striking the Earth within the next few decades.
Events Meriting Concern	2	A somewhat close, but not unusual, encounter. Collision is very unlikely.
	3	A close encounter, with 1% or greater chance of a collision capable of causing localized devastation.
	4	A close encounter, with 1% or greater chance of a collision capable of causing regional devastation.
Threatening Events	5	A close encounter, with a significant threat of a collision capable of causing regional devastation.
	6	A close encounter, with significant threat of a collision capable of causing a global catastrophe.
	7	A close encounter, with an extremely significant threat of a collision capable of causing a global catastrophe.
Certain Collisions	8	A collision capable of causing localized destruction. Such events occur somewhere on Earth between once per 50 yr and once per 1,000 yr.
	9	A collision capable of causing regional devastation. Such events occur between once per 1,000 yr and once per 100,000 yr.
	10	A collision capable of causing a global climatic catastrophe. Such events occur once per 100,000 yr, or less often.

Figure 22. Categories within the Torino Scale.

3. MISSION CONFIGURATIONS

In considering how to counter an incoming object, it is important to consider first whether it is better to push the object out of the way or to break it up into small pieces. Each option offers its own set of advantages and disadvantages. Additionally, one must consider how to deliver the energy needed to deflect or fragment the object. Three methods are discussed here: remote station, interception, and rendezvous. Under the remote station approach, no complex spacecraft would be sent out to the approaching NEO. Instead, all operations are conducted remotely, probably from the vicinity of the Earth, with beamed energy or projectiles being used to perform the deflection or fragmentation. A strategy based upon interception would involve sending spacecraft out on an intercept trajectory with the incoming NEO; the resulting high velocity impact(s) would accomplish either deflection or fragmentation. Rendezvous-based techniques are more propulsively demanding, as they require one to dispatch hardware to actually match orbits with the incoming NEO.

At first consideration, it would seem that the decision over deflection versus fragmentation is interlinked with the method chosen. An intercepting object would deliver all of its energy at once, tending to cause fragmentation instead of deflection. However, an incoming NEO could be deflected by a series of intercepting objects, each imparting enough momentum to slightly perturb its orbit without causing fragmentation. Similarly, it may seem improbable to actually rendezvous with an incoming object only to subsequently break it up. However, if there were a finite amount of time needed for the system to deliver the fragmentation energy, then rendezvous becomes necessary.

3.1 Deflection Versus Fragmentation

Figure 23 illustrates the concept of deflecting an incoming object away from an orbit that intersects with the Earth. In the case illustrated, it is anticipated that the deflection mechanism would require a significant period of the NEO orbit to deliver the energy necessary to perturb its orbit. Figure 23 shows the commencement of deflection before aphelion; if undeflected, the object would collide with the Earth near perihelion.

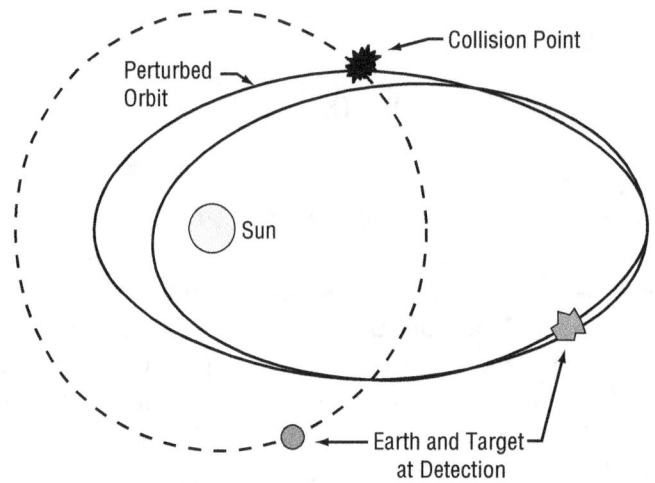

Figure 23. Illustration of deflection method of threat mitigation.

When considering strategies based on deflection, it is important to establish what level of perturbation is necessary to consider the Earth as being safe from collision. One might argue that "a miss is good as a mile," but some margin of error is necessary when designing a system that would deflect incoming objects. The literature commonly uses a figure of 3 Earth radii as a minimum safe approach distance for a deflected object. This value takes into account the uncertainty in astrodynamical constants that affects trajectory modeling accuracy for the incoming object.

Figure 24 illustrates the concept of fragmentation. At first sight, this might seem the best approach as the object's destruction means that it cannot threaten the Earth on a later orbit. In addition, there is no requirement to deliver the energy to the NEO in a distributed manner; thus, it can be defeated in one shot. Finally, as recent Hollywood blockbuster movies clearly demonstrate, there is a unique emotional satisfaction to be derived from destroying a life-threatening object in this emphatic manner.

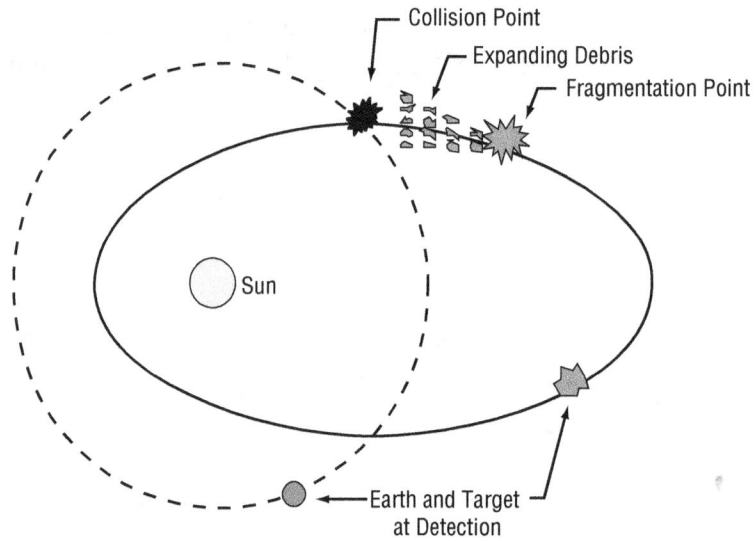

Figure 24. Illustration of fragmentation method of threat mitigation.

Despite its immediate tabloid appeal, fragmentation does introduce several issues that, on reflection, make it less attractive than deflection. It is important to break up the object into relatively small components. To break up the object into just a few pieces could actually exacerbate the damage to the Earth, with several distributed impacts occurring instead of one large impact. The fragmented pieces can "draft" off one another in the atmosphere; i.e., following pieces can travel within the slipstream of a leading piece and thus reach the ground relatively intact, alleviating burnup. For these reasons, the suggested fragmentation criteria is that no fragment should have a diameter >10 m. A major problem arises because asteroids and comets are suspected to have a very heterogeneous composition with significant internal structural flaws. Energy deposited into such objects cannot be expected to cause uniform fragmentation.

3.2 Remote Station Versus Interception Versus Rendezvous

The three modes considered in this study to deliver deflection or fragmentation energy to the incoming object are remote station, interception, and rendezvous. The remote station mode is depicted in figure 25. A station remains in orbit around either the Earth or the Sun. Energy can be delivered in the form of projectiles fired from the station by a mass driver or by a focused beam of coherent light, such as a solar lens or a laser. The advantage of such a system is that it remains close to the Earth and is easily maintained and upgraded. Also, the system can start deflecting the incoming object almost immediately—which might be months or years—during which an interceptor or rendezvous system would take to reach the object.

However, there are also several disadvantages. Targeting of the beam or stream of projectiles is not a trivial issue. For instance, targeting is required to within 1.4×10^{-5} to 2.8×10^{-6} arc s for objects between 1 and 5 au, the approximate orbital radii of Earth and Jupiter. Focusing the beam on the object across such vast distances would also be very challenging. There would be no vehicle in the vicinity of the object that could accurately assess the effect of the beam. Any such assessment would have to be conducted remotely from terrestrial- or Earth-orbiting platforms. This need for remote sensing over large distances would make it more difficult to properly assess the effect on the NEO. Moreover, unless the station is placed in a polar orbit, the object will almost certainly be eclipsed once per revolution. Polar orbits would require additional launches to deliver the station into orbit and would result in higher radiation exposure. Finally, a remote station would only be able to deflect incoming objects radially away from the station. Over a finite time period, the station and object will move relative to one another, causing the deflection vector to rotate, thereby wasting some of the beamed energy. Also, one must remember that radial deflection may not be the most efficient deflection strategy. See section 6 for a discussion of optimal deflection directions.

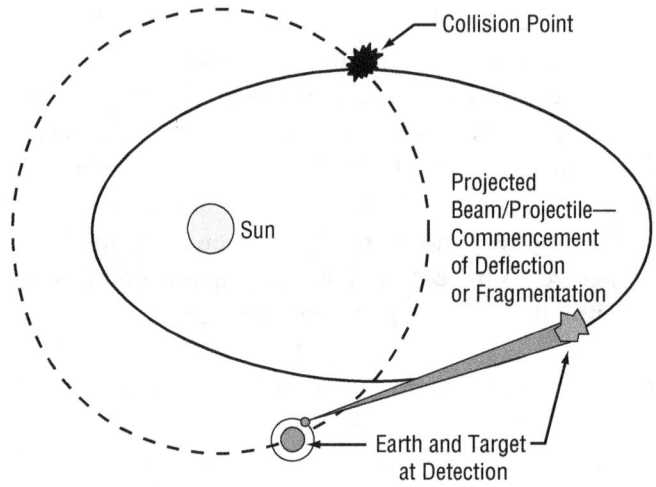

Figure 25. Delivering deflection or fragmentation energy by the remote station mode.

The interception strategy is depicted in figure 26. After NEO detection, the interceptor is deployed to intersect it later in its orbit. At this point, deflection or fragmentation can commence. In most cases, the interceptor will have substantial kinetic energy relative to the NEO. Thus, the interception option allows use of some of the energy initially stored in the outbound propulsion system to be delivered to the NEO. Interception options tend to be relatively simple, capitalizing on the high kinetic energy that is naturally available. The propulsive requirements for interception are substantially less than for rendezvous. This difference is further discussed in section 6.

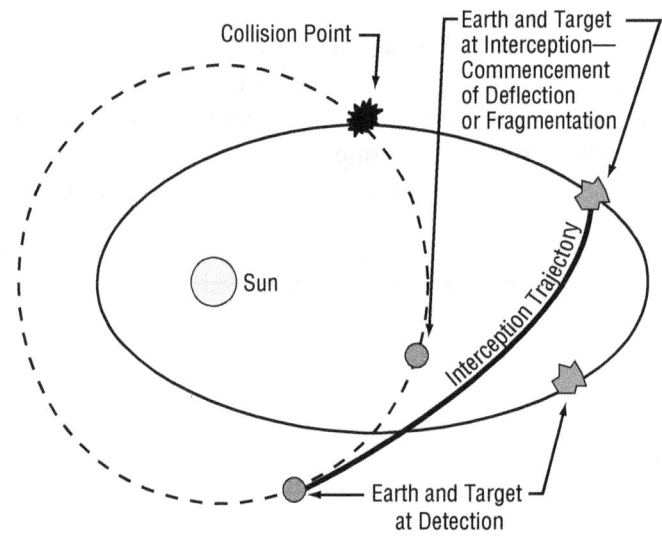

Figure 26. Delivering deflection or fragmentation energy by the interception mode.

Despite these advantages, the interception strategy also has its problems. The rate of closure between the interceptor and NEO can be as high as several tens of miles per second. This is an order of magnitude higher than the closure rate required for the kinetic kill vehicles used in the U.S. Global Missile Defense (GMD) program. The GMD program has had a mixed success rate in interception tests against simulated intercontinental ballistic missile (ICBM) warheads. While it is true that an incoming asteroid or comet will not maneuver to avoid destruction, and at 10 to 1,000 m in size, offers a larger target than a 3- to 5-m ICBM, the very high closure speeds still pose significant problems for guidance and terminal maneuvering.

Clearly, only a system with multiple interceptors dispatched in sequence, with the later vehicles capable of adjusting their trajectories, can provide the opportunity to continually sense changes in the NEO orbit. It should be noted that the interception strategy has the same type of deflection vector limitations as for the remote station strategy.

The final strategy is that of rendezvous with the incoming NEO, as shown in figure 27. After detection, the rendezvous system is deployed and matches orbits with the NEO later along its trajectory. This strategy is the one required for most types of deflection systems; it has several significant advantages. Targeting the NEO is much less difficult for a vehicle in a parallel orbit. This strategy offers the best opportunity to continuously evaluate the NEO during deflection or fragmentation operations. The limitations placed on the direction of the deflection vector that were encountered with the other two strategies are absent for the rendezvous option, allowing deflection in the direction that requires the least energy. Finally, this strategy has the greatest synergy with resource utilization missions.

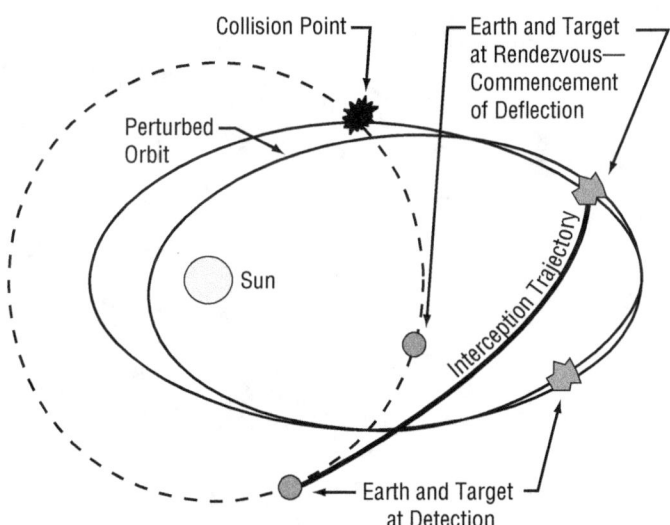

Figure 27. Delivering deflection or fragmentation energy by the rendezvous mode.

Of course, the propulsive requirements to rendezvous with an incoming NEO are much higher than that for interception. Additionally, the response time for a rendezvous system must include both the outbound rendezvous time and whatever inbound time is needed for the fragmentation or deflection process to take place. The rendezvous vehicle may be sufficiently distant from the Earth to make teleoperations difficult and would thus require significant onboard autonomy in an unknown environment that offers many opportunities for unexpected effects. Finally, during fragmentation or deflection, the rendezvous vehicle will probably be exposed to a hazardous environment filled with ejecta from the asteroid. The vehicle will have to be designed to resist this environment.

4. OUTBOUND PROPULSION

For both the interception and rendezvous techniques, neither fragmentation nor deflection can take place until the necessary system hardware is transported out to the approaching NEO. Some type of outbound propulsion system is required to accomplish this. Several propulsion systems are considered in this study. They were selected for their ability to meet the mission requirements, their level of technological maturity, and development status.

Each propulsion system has been assessed against a range of qualitative considerations that have been appropriately weighed by the MSFC study team. Although these considerations are of considerable merit in comparing the various outbound propulsion systems, being qualitative, they do not easily lend themselves to a numerical analysis. Instead, they are classified as high, medium, and low. When presented diagrammatically, they are indicated by the use of three colors—green, yellow, and red—to indicate decreasing levels of favor.

Each consideration is categorized as being either first or second order. First-order considerations include thrust level, scalability, long-term readiness, and compactness. High thrust levels are favorable because they reduce outbound trip times and gravity losses. Scalability is the parameter that denotes the ability of a proposed propulsion system to fulfill a range of propulsive needs, and to thus handle a variety of threat sizes. Since the vehicle will probably have to be constructed and kept until a threat is detected, long-term storage at a high state of readiness is a major issue. For similar reasons, compactness is also important. These considerations are all regarded as being of first order, as they directly affect the ability of the proposed system to meet the mission requirements.

Second-order considerations include usefulness of the system as a weapon, perceived safety, synergy with other NASA missions, and cost. The propulsion options under consideration are—of necessity—all high-energy systems that could cause extensive damage if misused. However, the category that assesses a system's usefulness as a weapon must consider whether use of the system as a weapon will outperform existing weapon systems.

NASA always considers safety to be an issue of primary importance. However, some high-energy systems are considered to be more threatening than others; i.e., nuclear systems. In these cases, extra effort must be expended to overcome public opinion obstacles, usually generated by the lack of public understanding and consequent mistrust of these specific technologies. The benefits of synergy with other NASA missions need no further explanation. For present purposes, the synergy consideration is divided into manned and robotic exploration missions as well as missions in which the use of asteroid and comet resources is a primary goal. Finally, development and deployment costs constitute the final consideration. These considerations are all regarded as being of second order as they do not affect the ability to meet the immediate mission objectives, but they do address political and economic issues involved in the deployment of a system.

4.1 Staged Chemical

Chemical systems are the most highly developed propulsion technologies currently available. They also offer the lowest performance of all the options considered for this project. In fact, chemical systems would not be able to handle many of the propulsive requirements for these missions without staging. For this reason only the high-performance liquid oxygen (lox)/liquid hydrogen (LH$_2$) propellant combination is considered. A two-stage lox/LH$_2$ vehicle is shown in figure 28. In the analysis that follows, a value of 465 is used for the specific impulse (I_{sp}) for each stage.

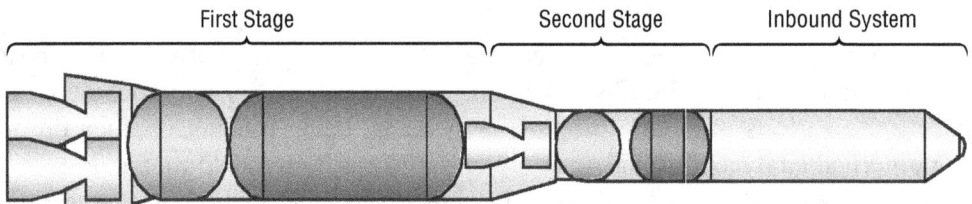

Figure 28. Two-stage lox/LH$_2$ vehicle. Image produced by INTROS.[12]

The analysis method used in this project is covered in detail elsewhere[13] and is only summarized here. Staging calculations requires knowledge of both the inert and propellant masses for each stage as well as the I_{sp} of each stage. The inert mass fraction is instrumental in calculating these masses; it is defined as

$$\varepsilon_n = \frac{m_{s_n}}{m_{p_n} + m_{s_n}} \ . \tag{1}$$

The inert mass fraction for the nth stage is the structural mass of the nth stage, m_{s_n} divided by the structural and propellant mass for the nth stage, m_{p_n}.

Predicting the inert mass fraction requires knowledge of historical vehicle designs. Curve fits using these historical data can then be developed (see fig. 29). Here, the inert mass fractions of historical launch vehicle stages listed in table 6 have been plotted against the amount of propellant contained in each stage. The line plotted represents a regression fit of the data. This regression line is given by

$$\varepsilon_n = 0.422 \log_{10}(m_p) - 0.0579 \ , \tag{2}$$

where propellant mass is in lbm. The Shuttle's external tank was not used to calculate this regression as its mass does not include a propulsion system. If included, this discrepancy would unfairly bias the results.

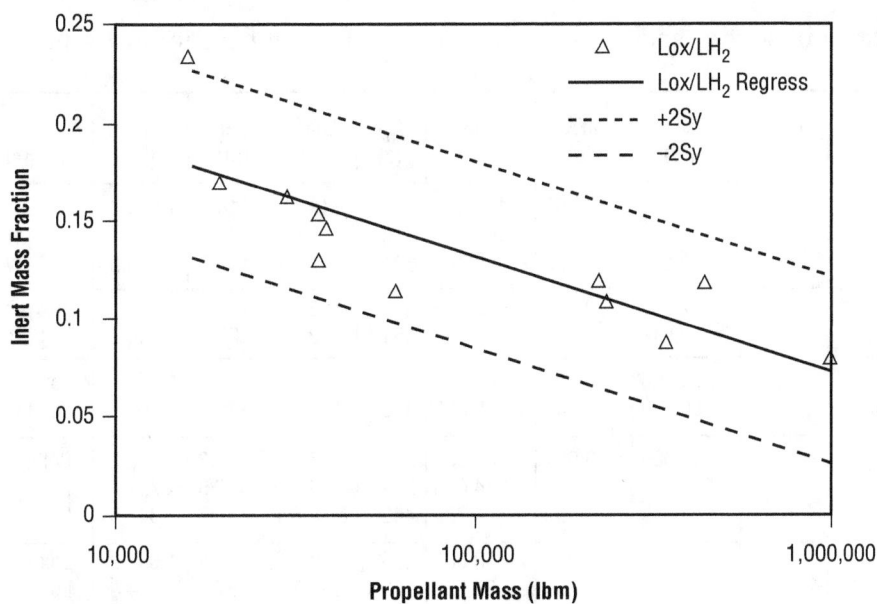

Figure 29. Regression curve fit of lox/LH$_2$ launch vehicle stages.

Table 6. Data on in-service and historical launch vehicles that have lox-/LH$_2$-powered stages.

Vehicle	Stage[1]	M_i^1 (lb)	M_f^2 (lb)	Thrust[3] (lbf)	I_{sp}^3 (s)	Dia[4] (in)	Span[4] (in)	Length (in)	Propellants
Ariane 5 (ESA)	0	593,043	74,957	1,455,049	259	118.1	118.1	1,220.5	Solid
	1	374,785	33,069	241,670	310	212.6	212.6	1,181.1	Lox/LH$_2$
	2	20,679	4,806	6,173	324	157.5	157.5	129.9	N$_2$O$_4$/MMH
Atlas IIIb (USA)	1	432,300	30,200	932,669	311	122	118.1	1,141.7	Lox/RP–1
	2	50,618	4,696	44,584	451	122	122	460.6	Lox/LH$_2$
Atlas V Centaur III	1	673,300	46,060	860,200	311.3	150	150	1,278	Lox/RP–1
	2	50,620	4,700	22,300	450.5	120	120	462	Lox/LH$_2$
Delta III (USA)	0	42,609	5,031	141,250	273	47.2	47.2	578.7	Solid
	1	230,112	15,040	244,096	254	94.5	157.5	787.4	Lox/RP–1
	2	42,060	5,459	24,736	462	94.5	157.5	346.5	Lox/LH$_2$
Delta IV	1	499,000	9,000	650,000	365	200.4	200.4	1,606.8	Lox/LH$_2$
	2	67,700	7,700	24,750	462.4	157.2	157.2	474	Lox/LH$_2$
H-1 (Japan)	0	9,753	1,532	58,206	232	31.5	31.5	236.2	Solid
	1	189,156	9,700	194,844	253	94.5	94.5	866.1	Lox/RP–1
	2	23,369	3,968	23,149	450	98.4	98.4	405.5	Lox/LH$_2$
	3	4,850	794	17,416	291	51.2	51.2	90.6	Solid
H-IIA (Japan)	1	250,500	30,000	191,000	338	157.2	157.2	1,464	Lox/LH$_2$
	2	43,200	6,600	30,800	447	157.2	157.2	362.4	Lox/LH$_2$
Long March 3D (China)	0	90,389	6,614	183,508	259	90.6	90.6	629.9	N$_2$O$_4$/UMDH
	1	394,627	19,842	734,033	259	133.9	275.6	909.4	N$_2$O$_4$/UMDH
	2	87,193	8,818	186,817	260	133.9	133.9	409.4	N$_2$O$_4$/UMDH
	3	45,415	6,614	35,274	440	118.1	118.1	346.5	Lox/LH$_2$
Saturn II	2	1,079,800	86,000	1,150,000	425	396	396	978	Lox/LH$_2$
Saturn IB	1	987,140	91,520	1,640,000	232	256.8	256.8	963.6	Lox/RP–1
Saturn IVB	3	261,400	28,380	200,000	426	260.4	260.4	711.6	Lox/LH$_2$
Saturn 1C	1	4,872,000	288,000	7,760,000	264	396	396	1,656	Lox/RP–1
STS (USA) RSRM	0	1,299,998	190,001	2,589,796	337	145.7	200.8	1,515.7	Solid
ET	1	1,655,615	65,984	0	363	342.5	342.5	1,846.5	Lox/LH$_2$
Orbiter	2	218,958	218,515	1,536,411	455	192.9	937	1,464.6	Lox/LH$_2$
Titan (USA)	0	498,758	74,512	1,314,999	238	122	122	1,019.7	Solid
	1	256,999	12,000	526,000	250	122	122	878	N$_2$O$_4$/Aerozine-50[6]
	2	64,348	5,849	101,999	316	122	122	311	N$_2$O$_4$/Aerozine-50[6]
	3	35,843	5,800	29,500	444	122	122	378	Lox/LH$_2$

[1] Vehicle families use multiples of particular stages to achieve a variety of performance requirements. Zero and first stages especially are frequently used in numbers to boost performance. The data listed represents one stage only; i.e., zero stage on the STS represents one RSRM, not two. Zero stage indicates strap-on boosters.
[2] Neither initial or final mass includes payload.
[3] Thrust and I_{sp} represent sea level values for all stages except the last. Vacuum thrust and I_{sp} are listed for the final stage.
[4] Dia is the diameter of the vehicle fuselage. Span is the total span of the vehicle including fins and wings, if any.
[5] There are no engines on the Space Shuttle external tank.
[6] Aerozine-50 is a 1:1 mixture of UDMH and N$_2$H$_4$.

The energy required to deliver the threat mitigation system is defined as ΔV. ΔV is the difference between the velocity required for the final orbit and the velocity in the initial orbit. This requirement is defined by the trajectory and is discussed further in section 6. Each stage produces a portion of the total ΔV requirement. There is an optimal distribution between the stages that is defined by

$$\Delta V = \sum_{n=1}^{N} g_0 I_{sp_n} \frac{\alpha g_0 I_{sp_n} + 1}{\alpha \varepsilon_n g_0 I_{sp_n}}, \tag{3}$$

where g_0 is the gravitational constant at the Earth's surface, 9.8066 m/s², I_{sp_n} is the specific impulse of the nth stage, and α is a Lagrange multiplier.

These values can be broken into stage masses using the following equation:

$$\Delta V = \sum_{n=1}^{N} g_0 I_{sp_n} \ln\left(\frac{M_{i_n} + m_{pay_n}}{M_{i_n} - m_{p_n} + m_{pay_n}}\right). \tag{4}$$

M_{i_n} is the stage mass or the sum of M_{s_n} and m_{p_n}, and m_{pay_n} is the mass of everything above the nth stage. Thus, for the final stage, m_{pay_n} is the payload or the mass of the threat mitigation system. For the $(n-1)$th stage, the mass is the payload mass together with the total mass of the nth stage. Finally,

$$MR_n = \frac{\alpha g_0 I_{sp_n} + 1}{\alpha \varepsilon_n g_0 I_{sp_n}} \tag{5}$$

$$\lambda_n = \frac{1 - \varepsilon_n MR_n}{MR_n - 1}, \tag{6}$$

where MR_n and λ_n are, respectively, the mass ratio and payload fraction for the nth stage. Payload fraction is also defined in a manner similar to inert mass fraction:

$$\lambda_n = \frac{m_{pay_n}}{m_{p_n} + m_{s_n}}. \tag{7}$$

Finally, mass ratio is defined as

$$MR_n = \frac{M_{i_n}}{M_{f_n}}. \tag{8}$$

The method for solving these equations is as follows: First, the inert mass fractions are estimated for each stage. Our code used a value of 0.2 for each stage. Then, equation (3) can be solved for the Lagrange multiplier. Next, equations (6) and (5) can be solved in series. Equations (7) and (8) can then be used to derive propellant and inert masses. The propellant masses are used in equation (2) to calculate new inert mass fractions. This calculational scheme is repeated until convergence is achieved. Our code also executed this loop for a five-stage system down to a single stage to determine the least number of stages necessary to meet the ΔV requirements.

The Lagrange multiplier method is usually of limited use, as it does not handle ΔV requirements that include losses, such as gravitational or drag losses. However, these losses are minimal for the types of system envisioned, where the vehicle would be deployed in low-Earth orbit (LEO). That raises an operational concern; lox/LH$_2$ systems are not considered viable candidates for applications where long storage times are necessary. The systems envisioned would be applied in a situation where the vehicle would be assembled in advance and then placed in a parking orbit. Only when a threat is detected would the system be activated. In this type of application, the propellant tanks will be exposed to temperature cycling that will probably exacerbate the normal problems of cryogenic propellant storage. However, the data used to predict inert mass fractions were for launch vehicles, which must handle high stresses during ascent and produce an initial thrust-to-weight ratio of 0.8–0.9 for upper stages and 1.2–1.5 for first stages. Both of these conditions are alleviated for in-space vehicles, which will result in significant mass savings. Therefore, for this level of analysis, it is assumed that these mass savings will counteract the additional mass necessary for the active and passive thermal protection systems necessary to retain cryogenic propellants indefinitely.

Table 7 lists the qualitative considerations for chemical propulsion. The thrust level and scalability for chemical systems are excellent. However, the leak possibility for cryogenic propellants and the thermal protection requirements combine to indicate that long-term readiness will be difficult. The relatively low performance of chemical systems also makes them massive and the low density of LH$_2$ means that they will be physically bulky.

Table 7. Qualitative considerations for outbound propulsion using chemically-powered rockets.

First-Order Qualitative Considerations	
Thrust level	High
Scalability	High
Long-term readiness	Low
Compactness	Low
Second-Order Qualitative Considerations	
Usefulness as weapon	Low
Perceived safety	High
Synergy with other NASA missions	
Manned missions	High
Robotic missions	Medium
Resource utilization missions	Medium
Costs	
Development	Low
Deployment	Medium

Chemical propulsion does not lend itself for use of these systems as a weapon. Due to their extensive use over the last half century, they are perceived to be very safe by the general public. Chemical propulsion has strong applicability for manned missions where low trip times are important. For robotic and resource utilization missions, the level of potential synergy for chemical propulsion is considered to be medium. Development costs should be low for chemical systems due to the significant heritage and well-understood technologies of these systems, but deployment costs are higher due to the size and maintainability issues.

4.2 Nuclear Thermal Rocket

The nuclear rocket option was included to give a high thrust option, similar to the chemical option but with improved performance. The I_{sp} used for nuclear thermal rocket (NTR) propulsion systems is 850. Although almost twice as high as that for the chemical option, it is not enough to preclude the necessity of staging.

NTR engines weigh a considerable amount more than comparable chemical engines because of the required reactor and shielding mass. Since no NTR vehicles have been developed, the historical approach that was employed to calculate masses for chemically powered stages cannot be used. Instead, data from previous NTR engine development programs have been used, and some basic assumptions to apply it to the same type of analysis that was used for chemical stages have been made. A schematic of the NTR is shown in figure 30.

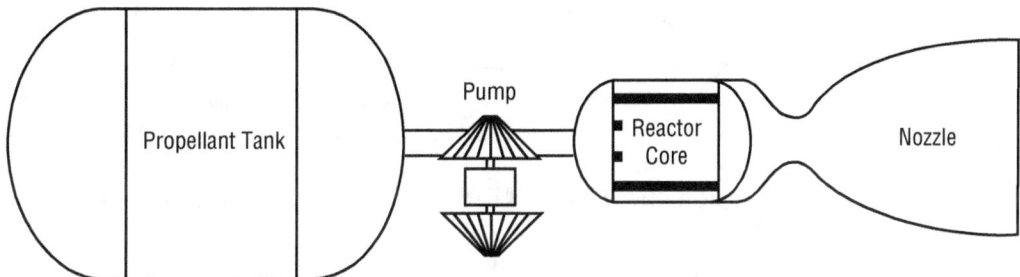

Figure 30. Schematic of a nuclear thermal rocket.[14]

Table 8 lists data from the Nuclear Engine for Rocket Vehicle Applications (NERVA) program. These data show a linear relationship between the total engine mass (m_{reac}) and thermal power produced by the reactor (P_{jet}):

$$P_{jet} = 0.127 m_{reac} - 53.8 \:. \tag{9}$$

Table 8. Data on nuclear rocket engines developed under the NERVA program.

	Mass (kg)	Thermal Power (MW)
Small engine	2,550	367
XE	7,700	1,140
NERVA	12,300	1,570
Phoebus-2A	41,679	5,320

This relationship is shown in figure 31.

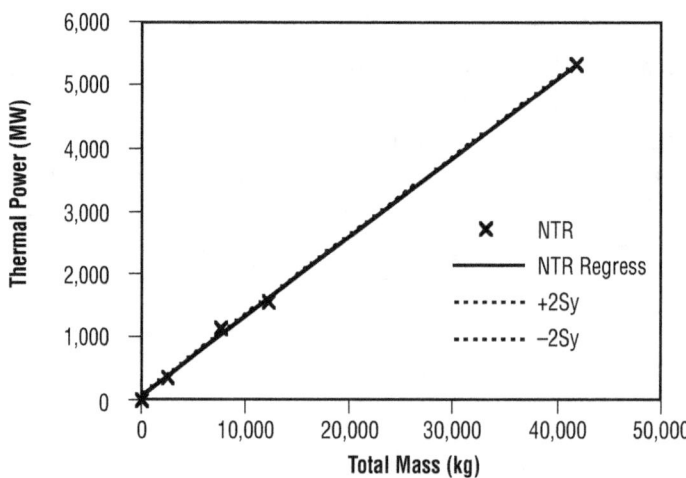

Figure 31. Regression curve fit of NERVA program-developed engines.

The power produced by the rocket jet is related to the I_{sp} and thrust produced by the rocket by the following equation:

$$P_{jet} = \frac{1}{2} g_0 I_{sp} F \ . \qquad (10)$$

For this analysis, it is assumed that the initial thrust-to-weight ratio for an NTR stage is 0.8. This is based to some extent on upperstage data from launch vehicles, but it represents a conservative estimate. Using this value, the power can then be related to the vehicle mass as follows:

$$P_{jet} = \frac{0.8}{2} g_0 I_{sp} \sum_{n=N}^{n} M_{in} \ , \qquad (11)$$

where the summation is intended to calculate the cumulative mass of the nth stage and all of the stages above it. The reactor mass can then be found by the solution of equation (9).

It would be convenient if this analysis could be conducted in the same way as for the chemical propulsion system from section 4.1. In this analysis, the reactor mass was divided by the stage mass and included with half of the inert mass fraction to yield an NTR inert mass fraction:

$$\varepsilon(\text{NTR})_n = \frac{m_{reac_n}}{M_{i_n}} + \frac{\varepsilon(chem)_n}{2} \quad . \tag{12}$$

This value can then be incorporated in the same analysis scheme as was used for chemical systems. The justification for this value is as follows. The reactor mass term accounts for the added mass associated with NTR engines. The chemical inert mass fraction accounts for the tanks, piping, thrust structure, and ancillary components necessary to make up a stage. It is divided by 2 because the NTR vehicle uses only one propellant, LH_2, instead of two. This assumption is optimistic because the LH_2 tank is much larger and heavier than the lox tank in a chemical vehicle. Additionally, an on-orbit system maintained indefinitely will require active and passive thermal protection, as did the chemical systems. However, the chemical inert mass fraction still contains the mass of the propulsion system too, which is superfluous here. Therefore, for the current level of analysis, it is assumed that these inconsistencies cancel each other.

Table 9 lists the qualitative considerations for NTRs. Like chemical systems, thrust level and scalability is believed to be high. Also, like chemical systems, maintenance of LH_2 propellant will make long-term readiness difficult. However, this system will be more compact than an alternative chemical system.

Table 9. Qualitative considerations for outbound propulsion using nuclear thermal rockets.

First-Order Qualitative Considerations	
Thrust level	High
Scalability	High
Long-term readiness	Low
Compactness	Medium
Second-Order Qualitative Considerations	
Usefulness as weapon	Low
Perceived safety	Low
Synergy with other NASA missions	
Manned missions	High
Robotic missions	Medium
Resource utilization missions	High
Costs	
Development	Medium
Deployment	Low

An NTR system would have little usefulness as a weapon. However, as a nuclear system, its perceived safety to the general public is rather low. This system offers high synergy for manned and resource utilization missions. However, these systems are most efficient at larger sizes; thus, they will be less effective for smaller robotic missions. Development costs are complicated by the fact that testing nuclear systems introduces a number of environmental issues. Deployment costs should be favorable due to their similarity with chemical systems, but with greater compactness.

4.3 Nuclear Pulse

The nuclear pulse concept was first considered only as an asteroid deflection technique. However, transport of the deflection system out to the NEO needs to be accomplished as swiftly as possible, and among propulsion techniques using known technology, external pulsed plasma propulsion (EPPP) or nuclear-pulsed systems are the best possible performers. EPPP utilizes not only the fission energy liberated in a nuclear device but is also substantially enhanced by a fusion energy release. Performance approaches that of a fusion-driven spacecraft, but without the additional challenges inherent in fusion technology.[15] As the fission fuel fragments ejected are extremely massive nuclei, thrust is also considerably higher than for pure fusion concepts.

There are several concepts that have been proposed for EPPP, including the standard pusher plate,[16] rotating cable pusher,[17] pusher plate,[18] magnetic field,[19,20] and large lightweight sail/spinnaker (Medusa concept) suggested by Solem.[21] All of these momentum transfer mechanisms (MTMs) utilize the same nuclear detonation energy source. Each couples the tremendous burst of high-velocity particles to the spacecraft by spreading the intense shock over a longer and more tolerable time period.

In this preliminary study, the most conventional—a term loosely applied to pulsed nuclear rockets—approach was taken: the pusher plate and shock absorber configuration. The specific geometry and scaling model was primarily derived from the original Air Force Program, ORION (fig. 32).[22] This classified program began in 1958 and ended in 1965, at a cost of about $8 to $10 million—significant research funding for those years. This program achieved tremendous technical success, but failed to maintain political support for a number of reasons. Significant data have recently been declassified and served as the basis for the generic vehicle calculations presented here. In 1999, a smaller NASA study, Project Gabriel, assessed the ORION concept, as applied to a smaller vehicle design, using current materials and technologies (fig. 33).[23,24] A significant portion of that work was also used in this study. The pulse unit designs are extremely general in nature due to the sensitive nature of the technology involved. However, the results appear to be realistic and well within current state of the art for such devices.

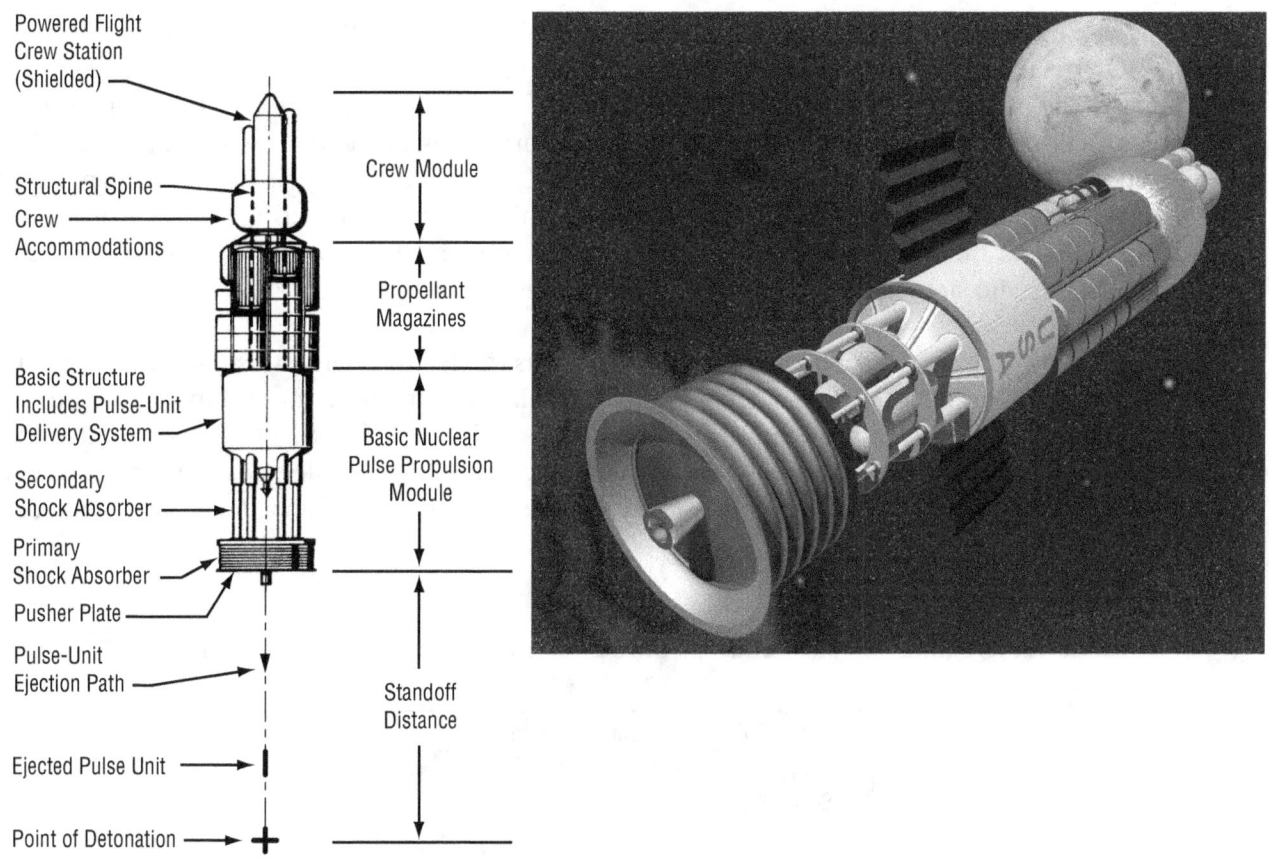

Figure 32. U.S. Air Force ORION spacecraft (1964).

Figure 33. NASA Gabriel spacecraft (1999).

47

The vehicle mass estimator uses geometry and density parameters that are based on rough scaling of the ORION vehicle. A 30-percent mass contingency factor was applied across the entire estimated vehicle mass. The pusher plate was assumed to be constructed from solid titanium; the primary shock absorber was a three-tiered carbon structure and the secondary shock absorber consisted of a set of long, gas strut type systems. A spreadsheet analysis was developed, which included gross estimates on pulse unit volume, packing fraction, and number of levels required. Although an effort was made to ensure enough spacing to preclude a thermal reactor critical geometry being created out of the pulse unit storage, no calculation or neutronic simulation was used to check for this issue.

Evaluation of the first-order qualitative considerations (table 10) clearly illustrates the outstanding technical merit of this system. It is the only system that has all high (green) ratings for the primary parameters. It also has excellent synergy with many other NASA missions. The robotic mission category is labeled a medium, since nuclear pulse has less potential application for smaller, near-term probes, but it receives high marks for interstellar type missions, sample returns, multiplanet tours, and large-scale robotic exploration.

Table 10. Qualitative considerations for outbound propulsion using nuclear pulse.

First-Order Qualitative Considerations	
Thrust level	High
Scalability	High
Long-term readiness	High
Compactness	High
Second-Order Qualitative Considerations	
Usefulness as weapon	Medium
Perceived safety	Low
Synergy with other NASA missions	
Manned missions	High
Robotic missions	Medium
Resource utilization missions	High
Costs	
Development	High
Deployment	High

The second-order qualitative considerations, which are negative, result from concerns over security of nuclear material and technical knowledge. The high cost is primarily generated by the significant need to regulate and secure fissionable material and prevent nuclear weapon proliferation. The fissionable material is readily available, and using it in peaceful space applications actually achieves cost savings in other areas. There are tons of highly enriched uranium and plutonium available from decommissioned Cold War weapons, both in the United States and in Russia. This material is presently being stored at a high cost to both nations and reprocessing it for burial will be extremely expensive. Furthermore, no matter what is done to this material, it will always be possible, although difficult, for someone to extract and reprocess it for weapons. Only in space is it safe from theft. When used as rocket propellant, it is rendered completely unusable in any future application. Not only is the exhaust material spread out into space, but much of the material will escape the solar system. The other factors responsible for the high cost result from the large size of the vehicle and also from testing issues. Pulse unit tests will require underground testing as well as several deep-space tests. The EPPP vehicle will require a monolithic pusher plate of 5–10 m (8 m was used in the final design developed here), some shock absorber development and testing, and a very fast pulse unit dispenser system. The initial launch costs could also be high, although comparable to other options considered in this study.

Other negative considerations also arise from concerns over security. Perceived safety is assessed as low, since this type of propulsion system is similar to a weapon system or could even be used as one. Although a pulse unit would be destructive if accidentally or purposely detonated on Earth, it would cause rather limited damage, since the devices are very small and not designed to generate the weapon effects of an atomic bomb. Note that the device will be only meters away from the spacecraft, and its power is held to a minimum, since many hundreds of these "pulses" must be endured. The arming can be made fail-safe and only possible in space/onboard the spacecraft.

Nations that have large quantities of fissionable material can easily build large weapons and those who would have only small amounts, such as a terrorist-supporting state, will find it nearly impossible to assemble, much less design and fabricate, the many extremely precise mechanisms that are required to make small amounts attain supercriticality. As far as security and safety are concerned, the historical record is favorable. No nuclear weapon has ever been lost or accidentally detonated, even after accidents in which a device has been inadvertently dropped from an aircraft. This record has been maintained for over 50 yr with thousands of weapons being moved around the world. It would seem very simple to build and store the devices for only several months at one U.S. location before shipment into space. It is apparent that all of these many mitigating points are not likely to counteract the likely perceived political risk. Like the original ORION program, this concept suffers from a basic human fear of ourselves, and it will be difficult to overcome this, even for purposes of asteroid defense.

The calculations were based on two external inputs: mission ΔV and payload mass. The mission ΔV was based on the trajectory analysis and is discussed in other sections. The payload mass is based on the nuclear pulse deflection option, it being logical to assume that if approval to build and launch the EPPP vehicle (fig. 34) was obtained, the asteroid threat mitigation could then be accomplished with the same type of technology. This conclusion is not absolutely certain, as there may be some chance of using another deflection scheme, or more likely, that a chemical or electric system would be used for outbound propulsion, but carrying a payload of nuclear devices designed to deflect the asteroid.

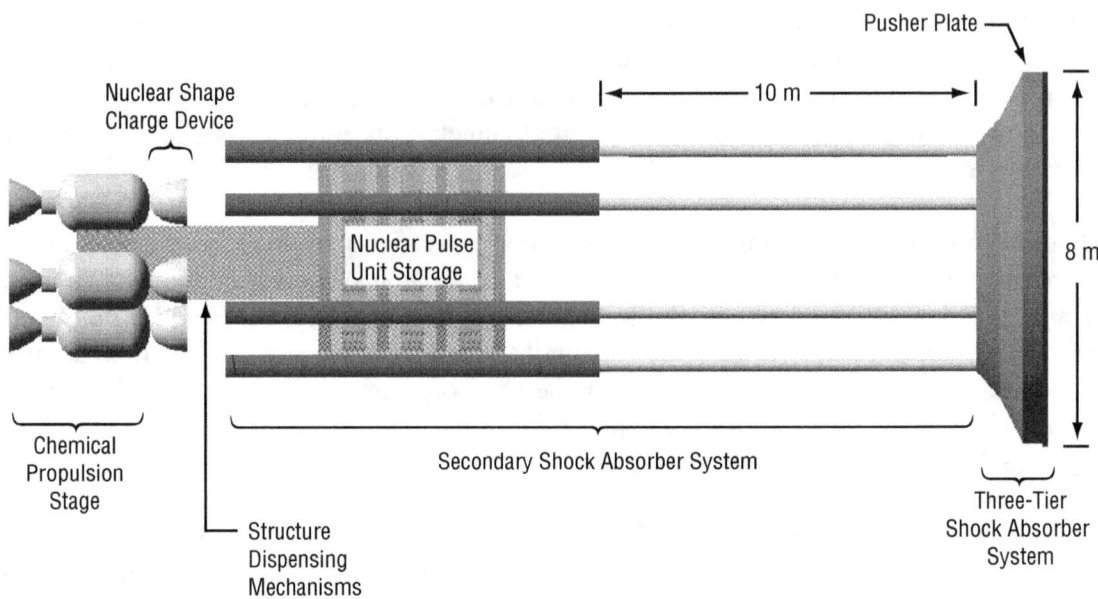

Figure 34. EPPP concept vehicle.

The calculations then proceed with two basic subsystem designs: (1) The vehicle's MTM, which is the pusher plate and the shock absorber system, and (2) the pulse unit design. From these two independent calculations, and using the required mission ΔV, an iterative solution is employed to determine the propellant required for the mission (the propellant being the number of individual nuclear devices or pulse units needed). Also, there is structure mass and associated volume that scales with the number of pulse units required.

The MTM was based on simple geometry and density considerations. The plate was a simple disk shape, although the real design would have a special taper profile as developed during the ORION program, with thickness and diameter specified and a mass calculated using a density, assuming a homogenous material. In this point design, the diameter was 8 m with a thickness of 3 cm. The material selected was titanium, which has a mass density of $\approx 4,500$ kg/m^3.

The primary shock absorber was a three-tier block of carbon microfiber springs. The original ORION program specified gas-filled toroidal chambers, whose mass was similar to that of the carbon material, but which was not as safe or reliable. Again, a simple geometry and density approach was used. Each tier was 0.5 m thick and the first; i.e., that connected to the metallic pusher plate, was cylindrical in shape with the same diameter as the pusher plate. The next two tiers were cone shaped rather than cylindrical. On one side, each has the same diameter as the preceding tier, and on the other side, the diameter is three-quarters of that value. The volume is calculated for each section and the density is used to determine the mass. Since the shock absorber is envisioned to be leaf spring or coil structured, a large percentage of the volume was considered to consist of voids. For the carbon material with an assumed density of 1,600 kg/m^3, a carbon fraction of 10 percent was specified. This was assumed without any rigorous analysis, since the material manufacturing process and actual property values are presently based on speculation.

The secondary shock absorber system was chosen to consist of gas tube struts, similar to the ORION engineering, but modeled as a simple hollow cylinder. Again, simple geometry and density estimates were made instead of using masses taken from the ORION study. Although the ORION numbers were based on relatively detailed engineering drawings, the simple geometry involved made it easy to develop a parametric calculation, which was used to explore a wide trade space area; i.e., the effect of changes in the number and size of the shocks could be readily investigated. The material selected was titanium with a mass density of ≈4,500 kg/m^3 and a wall thickness of 2 cm. Each of the six tubes was 0.5 m in diameter and 10 m in length. The sum of the three major components that make up the MTM system gave the total vehicle dry mass. A 30-percent mass contingency factor was applied to the total vehicle mass. The operation of the shock absorber system is illustrated figure 35.

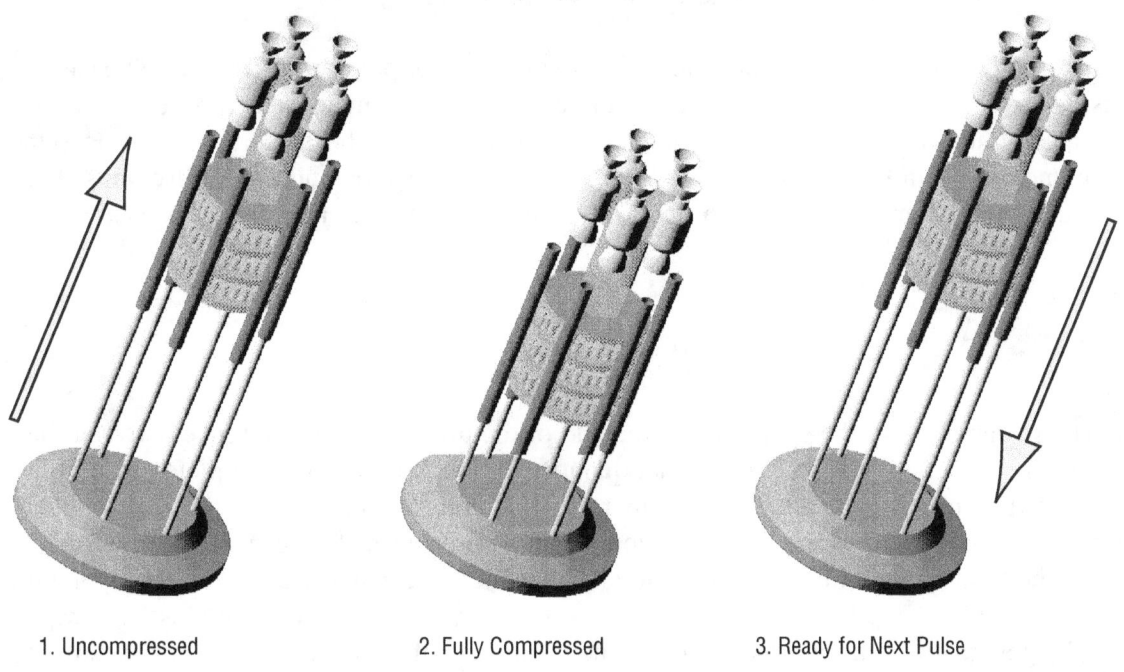

1. Uncompressed   2. Fully Compressed   3. Ready for Next Pulse

Figure 35. Shock absorber operations.

The pulse unit performance parameters were based upon the amount of energy released per nucleus fissioned, the mass of U235, and the inert mass. The inert mass includes everything else which comprises the complete device: impurities in the U235 fuel, high explosives, arming circuitry, columniation structure, channel filler, or low atomic number propellant and casing/mounts. The energy released per U235 nucleus fissioned is ≈185 MeV. This accounts for the thermal energy of the main fragments only and not the neutrons, product decays, neutrinos, or prompt gamma rays. This value was used to determine the theoretical average energy release per kilogram of fuel (7.592×10^{13} m^2/s^2), using Avogadro's number and a conversion factor. An estimated "burnup" fraction (percent of U235 nuclei actually split in the reaction) was assumed and the total energy released by the pulse unit was then determined; i.e., burnup fraction times mass of fuel in the pulse unit (kg) times the energy per kilogram. The chemical explosive energy required to start the nuclear reaction was neglected. The total fission energy was then assumed to be absorbed evenly into the total pulse unit mass and the average velocity of the particles determined from the equation:

$$E = 1/2 mv^2 \ . \qquad (13)$$

I_{sp} was determined from the average particle velocity times an effectivity factor, divided by the gravitational constant; i.e., $I_{sp}=v \times F_E/g$, where $g=9.807$ m/s^2). The effectivity factor (F_E) in the I_{sp} equation accounts for the collimation factor; i.e., plasma burst preferential direction, pusher plate diameter, and standoff distance (distance from pulse unit detonation to the center of the plate). The effectivity factor equation was derived by Thane Reynold.[16] From this I_{sp}, a total energy yield was determined and was expressed as an equivalent number of kilotons of tri-nitro-toluene (TNT).

The last spreadsheet calculation determines the number of pulse units needed to be carried as propellant using a spreadsheet goal seek iteration routine. An initial estimate of the propellant mass is made. This estimate is multiplied by a generic tankage fraction, assumed to be 15 percent. This accounts for all the storage, mounting, and dispensing mechanism hardware needed to handle the pulse units. The mass of the MTM and payload are then added to obtain the start mass of the spacecraft. Using the ideal rocket equation,

$$M_f = M_i e^{-\frac{\Delta V}{g_0 I_{sp}}} \ , \qquad (14)$$

a final mass is generated for the spacecraft at the end of the mission. The difference between that and the start mass is the propellant used. Dividing that quantity by the mass of one complete pulse unit (mass of U235 fuel and inert mass) established the number of pulse units (rounding up to the next whole value). A 5-percent contingency factor was added to account for misfires, trajectory errors, and other performance losses. With the final number of whole pulse units determined, the actual total mass of propellant is determined and compared to the initial estimate. If the estimate was not sufficiently close, a new value was automatically generated by the computer; i.e., the new value is halfway between the initial estimate and the calculated value, and the calculation repeated. The conversion criterion was an error better than 1×10^{15} between the two values.

Other calculations were performed to estimate the size of the propellant magazines. The diameter of the storage area was held to half that of the plate, and the pulse units were estimated by the density of U235, which has a mass density of 19,000 kg/m^3, and inert material; i.e., low atomic number with a mass density assumed to average 1,500 kg/m^3. A 10-percent void fraction was assumed to determine the gross volume. Using a length-to-width ratio of 2, the size and shape was ascertained for a single pulse unit. Required "floor space" was found by assuming a square storage geometry; i.e., packing cylinders, and adding an additional 25-percent fluff factor. The floor area required was divided by the area available based on the one-half plate diameter criteria to estimate the number of storage levels. Finally, a level height of twice the pulse unit length was assigned and the total propellant volume was determined. This was used to ascertain whether the propellant could be reasonably launched into LEO by conventional chemical rockets.

The EPPP calculation approach was simplistic, but nonetheless is considered to be realistic. Due to the sensitive nature of any details concerning nuclear devices and their effects, the inputs were intentionally conservative and were obtained from open literature sources. No inference should be made regarding any parameters relating to the true composition, size, geometry, or efficiency of any real nuclear device. Fusion energy, as well as the use of plutonium fuel, was not considered. If employed, they would be expected to enhance performance, although they each have their own drawbacks, most notably, their natural radioactive decay, cost, and availability. The vehicle was also designed using simple geometric and density-based calculations. No estimate of the effects of erosion, radiation, or shock was made for the pusher plate design. The 30-percent mass margin was considered adequate, but not necessarily conservative, for such a preliminary design. Because of the conservative assumptions used in pulse unit design, the MTM was not additionally burdened with a higher mass margin.

Many small trade studies were conducted in order to gain an understanding of how the entire system would function. The vehicle has excellent ability to absorb mass growth; a larger vehicle actually yields improved performance. Surprisingly, the final mission design was not very sensitive to the plate diameter as long as the plate diameter-to-standoff distance ratio was not altered significantly. The acceptable vehicle trade space was found to be very large. This gives confidence in the practicality of such a propulsion system, despite the large uncertainty in the design parameter inputs. Even doubling the payload mass increases the vehicle gross mass by <20 percent or ≈50 more pulse units. In general, pulse unit yields were in the 3 to 4 kton range (present devices are normally in the many megaton range) and the range of I_{sp} was 2,500 to 5,000 s. The general conclusion from the analysis is that this technology has excellent performance capability and that there are no known technical barriers.

4.4 Solar Sail

The solar sail offers unique capabilities for rendezvous with an incoming object. Interception is also a possibility but it would not be a first choice for threat mitigation systems depending upon kinetic energy. Solar sails are capable of substantial inclination changes that are difficult to achieve with other propulsion systems. Sufficiently large sails are also capable of moving sizable payloads over long distances. Also, since most NEOs are expected to occupy orbits within that of Jupiter, a solar sail would remain close enough to the Sun to maintain significant propulsive capability. The analysis method discussed here follows that given elsewhere.[25,26]

There are several solar sail configurations available. The three most popular are square sails, disk sails, and heliogyros. The difference between these types of sails relates mostly to consideration of structural design. Square sails use a square sail sheet supported by booms and lines. The heliogyro uses long, narrow sail blades that are each connected to a central hub. The blades are kept taut through rotation of the entire sail, including both the blades and the hub. A disk sail attempts to incorporate advantages of both square sail and heliogyro. It has a circular sail sheet that is kept taut through rotation. Previous studies have suggested that, in most cases, the square sail is slightly more attractive than the other options. The analysis presented here is based upon a square sail design (fig. 36).

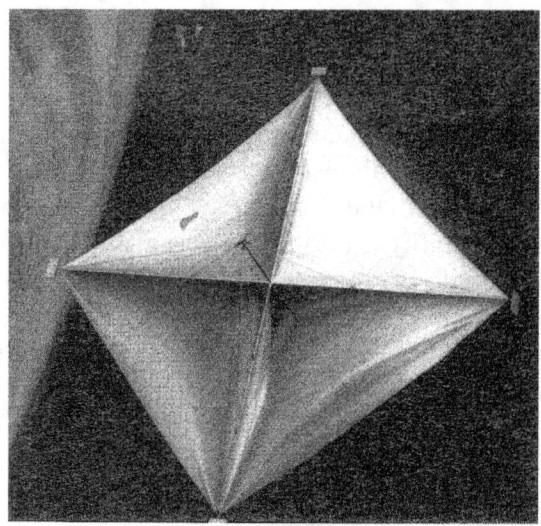

Figure 36. Artists concept of a billowing, square solar sail.[25]

The important dimensions for a square solar sail are illustrated in figure 37. Each side of the sail has a length (l). The total sail area (A_{sail}) is equal to the square of the sail length. The sail will normally be oriented at an angle of attack (α) relative to the plane perpendicular to the orbital radius vector. The sail will project an area (A) normal to the incoming luminous flux. The centerline angle (ϕ) is the angle between the sail normal and force vector. Similarly, the cone angle (θ) is the angle between the force vector and the incident luminous flux.

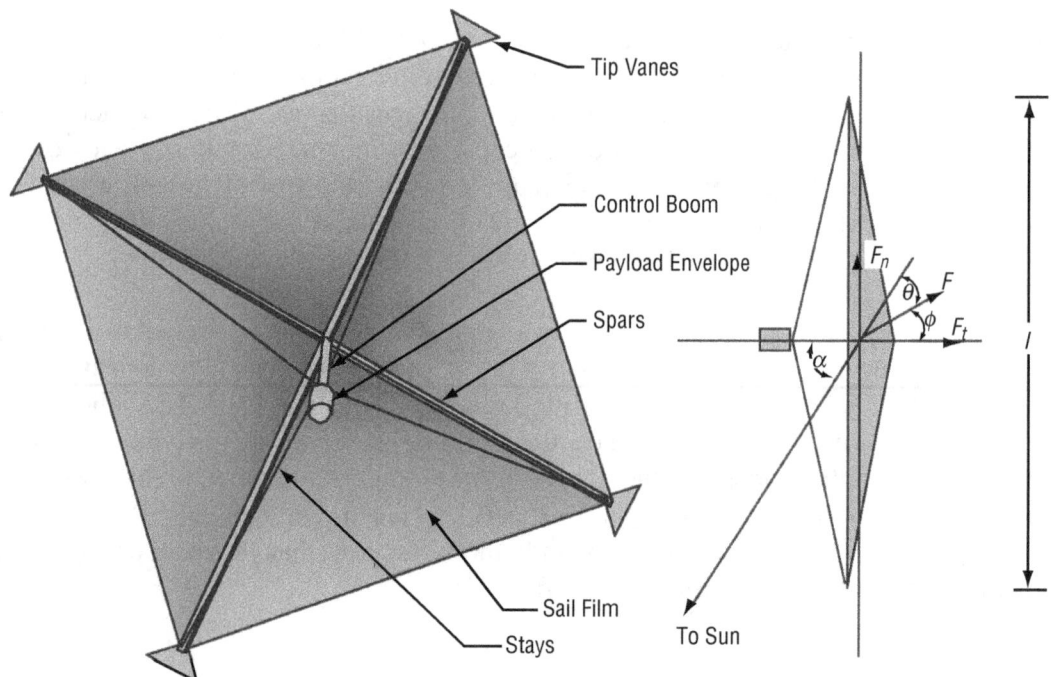

Figure 37. Schematic and dimensions for a square solar sail.

The first task in solar sail design is to model the luminous flux emanating from the Sun. The Sun should be modeled as a finite disk out to distances as far as 10 solar radii. The solar light pressure can be calculated as a function of distance from the Sun using the following equation:

$$P(r) = \frac{L_s}{6\pi c R_s^2}\left[1-\left(1-\frac{R_s^2}{r^2}\right)^{\frac{3}{2}}\right], \qquad (15)$$

where R_s is the radius of the Sun, r is the radius of the sail from the center of the Sun, c is the speed of light, and L_s is the solar luminosity.

The sail converts this pressure into thrust through absorption or reflection. Spectral reflection is the preferred method as it is the most efficient. However, 100-percent reflection is not easily achieved. Absorption of the incident light is less attractive because of the resulting thermal loads that must be radiated away from the reverse side of the sail. This emission process also produces its own radiation pressure, which partially cancels the motive force. The normal and tangential forces due to absorption are given as

$$F_{an} = PA\cos^2\alpha \qquad (16)$$

$$F_{at} = PA\cos\alpha\sin\alpha, \qquad (17)$$

where A is the cross-sectional area presented to the Sun. The tangential and normal forces due to reflection are calculated using

$$F_{rn} = PA\left(r's'\cos^2\alpha + B_f(1-s')r'\cos\alpha\right) \qquad (18)$$

$$F_{rt} = PAr's'\cos\alpha\sin\alpha, \qquad (19)$$

where r' and s' are the reflection coefficient and the spectral reflection coefficient, respectively. Also, B_f is the front non-Lambertian coefficient used to model the nonspectral reflection. The normal force due to emission is given as

$$F_e = PA(1-r')\frac{\varepsilon_f B_f - \varepsilon_b B_b}{\varepsilon_f + \varepsilon_b}\cos\alpha, \qquad (20)$$

where B_b is the non-Lambertian coefficient for the back side of the sail. Additionally, ε_f and ε_b are the emission coefficients for the front and back, respectively. The values used in this study for these optical coefficients can be found in table 11. These values can be traced back to a Jet Propulsion Laboratory (JPL) solar sail model developed during the 1970's to study a proposed Halley's Comet rendezvous mission.

Table 11. Optical, billowing force and other parameters used in solar sail analysis.[25]

Variable	Name	Ideal Value	Study Value
B_f	Front non-Lambertian coefficient	2/3	0.79
B_b	Back non-Lambertian coefficient	2/3	0.55
ε_f	Front emissivity coefficient	0	0.05
ε_b	Back emissivity coefficient	0	0.55
r	Reflection coefficient	1	0.88
s'	Specular reflection coefficient	1	0.94
C_1	Force coefficient 1	0.5	0.349
C_2	Force coefficient 2	0.5	0.662
C_3	Force coefficient 3	0	−0.011
σ	Sail loading parameter	Low as possible	6 gm/m²

The total sail normal and tangential forces can be determined by summing the above forces:

$$F_n = F_{an} + F_{rn} + F_e \tag{21}$$

$$F_t = F_{at} + F_{rt} \ . \tag{22}$$

The total force and centerline angle are determined using

$$F = \sqrt{F_n^2 + F_t^2} \tag{23}$$

$$\phi = \tan^{-1} \frac{F_t}{F_n} \ . \tag{24}$$

Finally, the sail temperature can be calculated by accounting for the thermal flux due to absorption and emission:

$$T = \left[\frac{(1-r')cP\cos\alpha}{\sigma'(\varepsilon_f + \varepsilon_b)} \right]^{\frac{1}{4}} , \tag{25}$$

where σ' is the Stefan-Boltzmann constant.

The sail can be allowed to billow under the luminous pressure and therefore will not present a plane surface. The cone angle under billowing can be determined by

$$\theta = \alpha - \phi \ , \tag{26}$$

and the total force adjusted for billowing is calculated using

$$F_{norm} = F(C_1 + C_2 \cos 2\theta + C_3 \cos 4\theta) \ . \tag{27}$$

The coefficients C_1, C_2, and C_3 are billowing force coefficients that were also determined during JPL's Halley's Comet study. These coefficients and the optical parameters are not universally applicable but are used here as a first approximation to take account of billowing and optical physics effects.

The mass of the sail can be calculated using

$$M_i = \sigma A + m_{pay} \ . \tag{28}$$

There are several figures of merit that are typically used to compare sail performance. First, the sail lightness number (β) is a characterization of the sail's acceleration compared to that due to the local gravitational force imposed by the Sun:

$$\beta = \frac{L_s}{2\pi G M_s c \sigma} \ . \tag{29}$$

Here, G is the universal gravitational constant, M_s is the mass of the Sun, and L_s is the solar luminosity.

Figure 38 illustrates the relevance of lightness number to sail orbital capabilities. For a $\beta < 1/2$, the sail travels on an elliptical trajectory around the Sun. For a β of between 1/2 and 1, the sail follows a hyperbolic escape trajectory. When $\beta = 1$, the force exerted on the sail exactly negates the gravitational force from the Sun. For values of $\beta > 1$, the sail trajectory is still hyperbolic, but with a thrust higher than the local gravitational force.

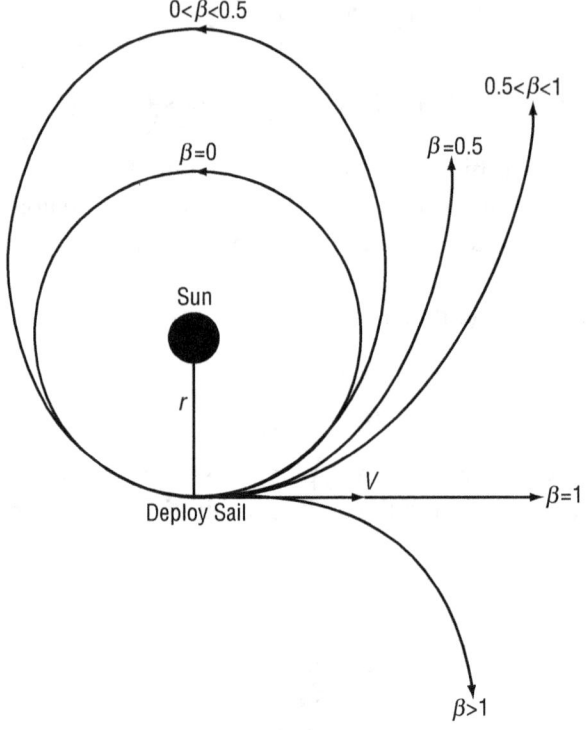

Figure 38. Sail trajectories relative to lightness number.[25]

The sail's efficiency is a measure of its ability to convert the incident luminous flux into a propulsive force; it is given by

$$\eta = \frac{F_{norm}}{2A_{sail}P} .$$ (30)

This value is a constant for any heliocentric distance or any sail area.

The characteristic acceleration is the acceleration that the sail will experience when it is normal to the Sun and at a distance of 1 au (the average radius of the Earth's orbit around the Sun). It is given by

$$a_0 = \frac{2\eta PA}{M_i} .$$ (31)

The sensitivities to several of these figures of merit are given by the following equations:

$$\frac{\Delta a_0}{a_0} = \frac{-1}{1+\frac{m_{pay}}{m_s}} \frac{\Delta \sigma_s}{\sigma_s}$$ (32)

58

$$\frac{a_0}{\Delta a_0} = \frac{-1}{1+\frac{m_s}{m_{pay}}} \frac{\Delta m_{pay}}{m_{pay}} \qquad (33)$$

$$\frac{a_0}{\Delta a_0} = \frac{1}{1+\frac{m_{pay}}{m_s}} \frac{\Delta A}{A} \; . \qquad (34)$$

These are the sensitivities to sail density, payload mass, and sail area, respectively.

Table 12 lists the qualitative considerations for solar sail outbound propulsion. Solar sails produce very low thrust. As even the least demanding missions require large sails, there is limited potential for scaling up the design. The sail does not have to be unfurled until a threat is identified; therefore, it can be maintained indefinitely in a very compact state.

Table 12. Qualitative considerations for outbound propulsion using solar sails.

First-Order Qualitative Considerations	
Thrust level	Low
Scalability	Low
Long-term readiness	High
Compactness	High
Second-Order Qualitative Considerations	
Usefulness as weapon	Low
Perceived safety	High
Synergy with other NASA missions	
Manned missions	Low
Robotic missions	High
Resource utilization missions	High
Costs	
Development	Medium
Deployment	Low

It is difficult to conceive of any way in which a solar sail could be used to cause harm. Sails have the ability to achieve a variety of mission types, as long as the total payload to be delivered is small. This makes it difficult to conceive manned missions using solar sails. The deployment of a furled sail presents few complications, although the unfurling process is not trivial. Sail development cost estimates are complicated by the difficulty in measuring the low propulsion levels on a ground-based facility.

Solar sails have unique abilities to change inclination and achieve non-Keplerian orbits. For these reasons, solar sails have proven in this study to have surprising capabilities to rendezvous with the varied orbits of potentially incoming objects. Intercept capabilities are less impressive, as sails do not accelerate quickly. However, the sail should be considered a strong contender for the outbound leg of any rendezvous concept.

4.5 Solar Collector

The solar collector is closely related to the solar sail; the principle is shown in figure 39. A solar sail with a curved configuration is used to focus incident light onto a secondary collector. The secondary collector directs the light outward so as to generate thrust. The curved sail centerline remains parallel to the vehicle Sun radius vector so that the sail generates the maximum possible thrust. This configuration avoids the loss of thrust experienced by a conventional solar sail when attempting to direct its thrust vector in an optimal direction. The focused light can also be effectively used to redirect incoming asteroids and comets, as will be discussed in section 5.

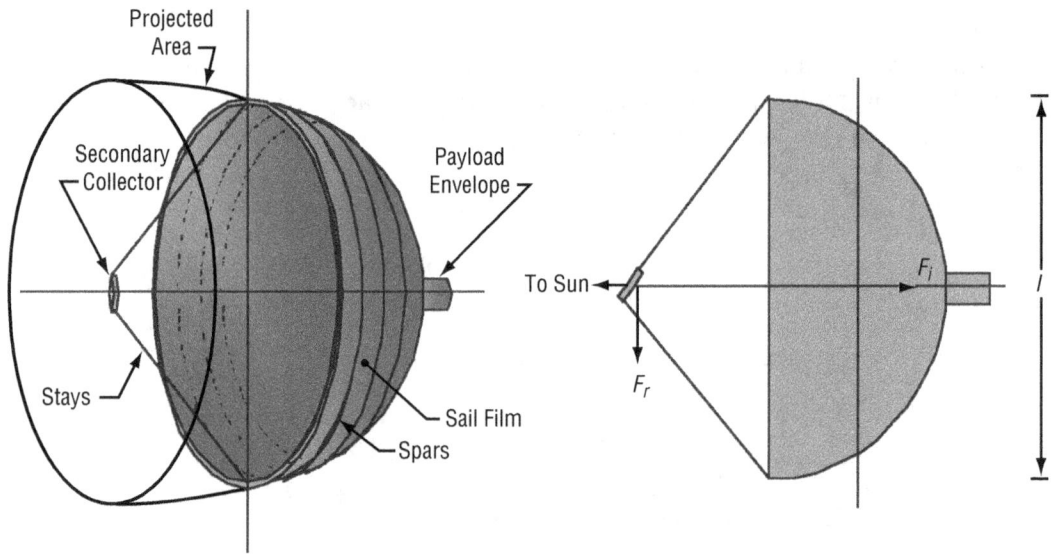

Figure 39. Solar collector configuration.

As shown in figure 39, the curved sail is assumed to be hemispherical, with the secondary collector located at the center of the projected sphere. Although this configuration is susceptible to chromatic aberration, this is not a concern for this application. Additionally, a hemispherical shape would limit the direction in which the secondary mirror could focus to the plane perpendicular to the collector centerline. Thus, figure 39 illustrates a sail with a projected hemispherical area, but with an actual area that is less. Accounting for the additional area of the hemisphere gives a conservative analysis. This is clear because it is obvious from the figure that the additional cross-sectional area capturing light is not significant. More efficient parabolic configurations may be found, but they would require ray tracing or other advanced calculations that are beyond the scope of this project. The captured light is a function of the cross-sectional area of the hemisphere. The sail mass is calculated using the hemispherical surface area and the sail-loading factor found in table 11. Use of these loading factors is somewhat optimistic, as a curved solar collector will require more structural support to maintain its shape than would a flat solar sail. However, this optimistic assumption is expected to negate the pessimistic hemispherical shape assumption. The solar collector uses the same optical parameters listed for the solar sail in table 11.

Design of the secondary collector is of particular importance for this concept. If the collector is made large, then the amount of sail area that it blocks may be prohibitive. The curved sail focus is found to be[27]

$$l' = \frac{1}{\frac{2}{l} + \frac{1}{r}} , \qquad (35)$$

where l is the diameter of the curved sail and r is the orbital radius of the sail from the Sun. For all realistic situations, l/r can be treated as being zero. For a 100-m-diameter sail, equation (35) yields a focus of 50 m, which gives the location of the secondary collector in figure 39. The magnification is shown to be

$$M = \frac{l'}{r} . \qquad (36)$$

The curved sail essentially produces an image of the Sun. The image size is

$$r_c = M \times R_s , \qquad (37)$$

where R_s is the radius of the Sun, 6.96×10^{10} m. For the sail >1 au, the magnification is 3.34×10^{-10} and the image size is 0.233 m. Therefore, the collector size need not be large and its shadow will be insignificant compared to the curved sail area projected normal to the Sun.

Table 13 lists the qualitative considerations for the solar collector. The solar collector is expected to have better thrust levels and scalability than the solar sail because of its more efficient use of the incident solar radiation. Note that the configuration cannot be folded as easily as can a solar sail. The collector has considerations similar to those of the solar sail in all other respects.

Table 13. Qualitative considerations for outbound propulsion using solar collectors.

First-Order Qualitative Considerations	
Thrust level	Medium
Scalability	Medium
Long-term readiness	High
Compactness	Medium
Second-Order Qualitative Considerations	
Usefulness as weapon	Low
Perceived safety	High
Synergy with other NASA missions	
Manned missions	Low
Robotic missions	High
Resource utilization missions	High
Costs	
Development	Medium
Deployment	Low

5. THREAT MITIGATION

5.1 Nuclear Fragmentation

The equations used to model the catastrophic fragmentation of a near-Earth solid body asteroid are based on the work of Thomas J. Ahrens, California Institute of Technology, and Alan W. Harris, JPL.[28] These equations are based on the assumption that an explosive device is placed deep enough below the asteroid's surface to produce near-optimum fragmentation. The location for optimum fragmentation is generally considered to be the geometric center of the target object.

In reference 28, starting with equation (65), p. 920, one can write

$$v(\text{cm}/\text{s}) = Hr' , \qquad (38)$$

where r' is the idealized radius of the spherical target body and H is a parameter which characterizes the hardness of the asteroidal material. For hard igneous terrestrial rocks, $H=5.72\times10^{10}$ s^{-1}. For soft terrestrial rock; e.g., sandstone, $H=2.90\times10^{10}$ s^{-1}.

An expression for shock wave internal energy per unit mass can also be written:

$$E = 0.5v^2 , \qquad (39)$$

where v is the shock-induced particle velocity and E is the energy. Fragmentation takes place when $E=E_{\text{fracture}} = \approx 10^{-7}$ erg/gm (energy density needed to break a 10-m object in two).

Finally,

$$r' = \frac{r}{W^{1/3}} , \qquad (40)$$

where r is the actual radius (in meters) and W is the required blast yield of the device. Combining these three equations gives an equation for the required blast yield as a function of asteroid radius in units of megatons of TNT:

$$W = \left(\frac{\sqrt{2E}2r10^3}{H}\right)^{3/2} , \qquad (41)$$

where the density of the target object is assumed to be 2 g/cm^3.

Figure 40 shows the blast yield required to fracture an asteroid into fragments smaller than 10 m in size. Although open to debate, it is generally assumed that fragments of this size would be much less likely than the original body to survive entry through Earth's atmosphere. Even if any fragments did reach the ground, the impact of these relatively small objects, spread over a large area, would be less damaging from a global point of view than from a single massive asteroid strike.

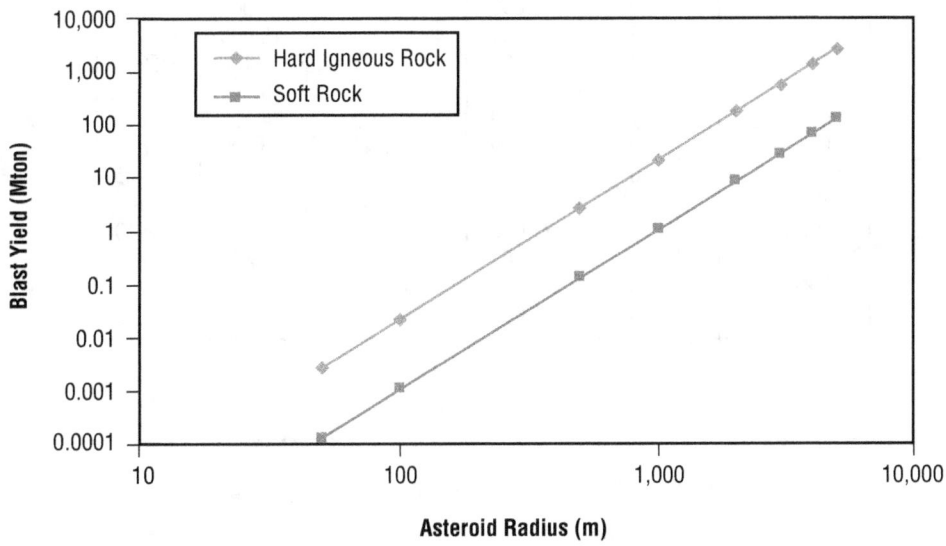

Figure 40. Blast yield—explosive placed at center of body—required for fragmentation as a function of asteroid radius.

These equations are admittedly somewhat ideal because they assume that:

(1) The asteroid is a perfectly spherical homogeneous structure.

(2) The explosive charge is placed at the exact geometric center.

(3) The explosion fractures the target body into pieces no larger than 10 m in diameter.

However idealized these assumptions may be, they do permit one to estimate the explosive power required for fragmentation to within an order of magnitude.

To gain a better understanding of the mass of the explosive payload to be delivered to the target body, data on existing nuclear warheads were tabulated using open literature sources in table 14 and graphed in figure 41.[29] As shown in the tabulated data, there are no existing devices that could catastrophically fragment an asteroid >2 km in diameter; this assumes that all of the ideal conditions listed above are satisfied.

Table 14. Nuclear device masses.

Designation	Yield (kton)	Mass (lbm)	Designation	Yield (kton)	Mass (lbm)
Mk–1	16	8,900	Mk–18	500	8,600
Mk–3	49	10,300	Mk–21	5,000	17,700
Mk–4	32	10,900	Mk–24	15,000	42,000
Mk–5	120	3,125	Mk–28	1,100	2,320
Mk–6	160	8,500	Mk–36	10,000	17,700
Mk–7	61	1,700	Mk–39	4,000	6,750
Mk–8	30	3,280	Mk–41	25,000	10,670
Mk–11	30	3,500	Mk–43	1,000	2,125
Mk–12	14	1,200	Mk–53	9,000	8,900
Mk–14	7,000	31,000	Mk–57	20	510
Mk–15	3,900	7,600	Mk–61	340	716
Mk–16	8,000	42,000	Mk–83	1,200	2,400
Mk–17	15,000	42,000			

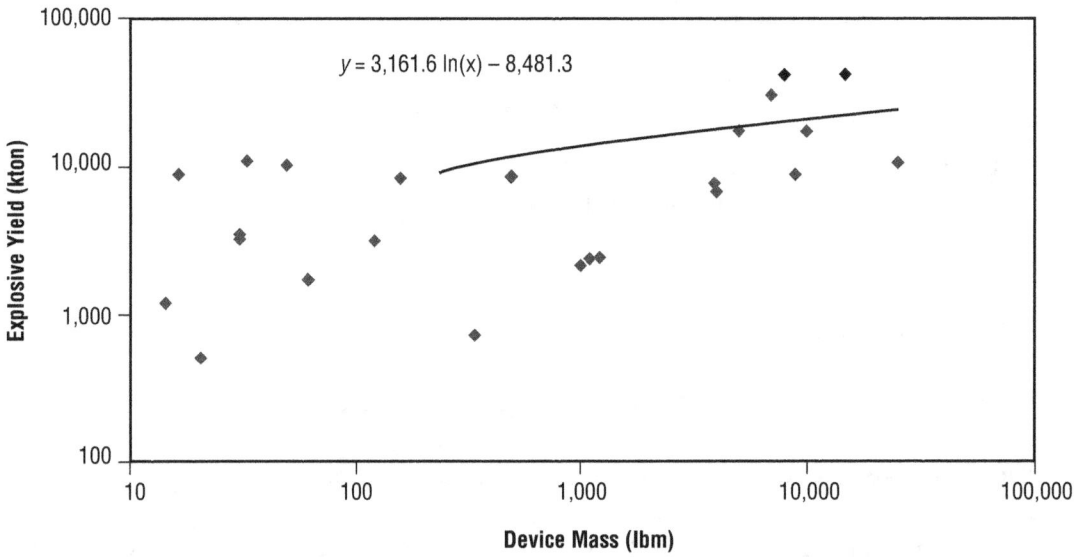

Figure 41. Device mass versus explosive yield.

The problem of exactly how the explosive device would be placed at the geometric center of the target body has not been addressed in this study. One idea considered is to utilize the same technology that is found in the "long-rod bunker buster" types of ordinance that the U.S. military employs against hardened underground facilities. This idea has the advantage of not requiring a ΔV breaking maneuver to rendezvous with and soft land on the target. Instead, the outbound kinetic energy is utilized to bury the device to the optimum depth. Again, using open literature sources, the physical characteristics of the BLU–113 penetrator bunker buster are listed in table 15.[30]

Table 15. BLU–113 penetrator characteristics.

Structure	Thick high-grade steel
Length	153 in
Diameter	14.5 in
Explosive	630 lbm (285 kg)
Overall Mass	4,400 lbm (2,000 kg)
Penetration of concrete	20 ft (≈6 m)
Fuse	FMU–143 series
Weapon system	GBU–28

These characteristics, especially the explosive mass-to-overall mass ratio, were used to estimate a representative total mass for the explosive payload that must be delivered to the target body (table 16). Now, the depth these first-generation penetration weapons can achieve has been explained by a simple rule of thumb. "For typical values for steel and concrete, we expect an upper bound to the penetration depth to be roughly 10 times the missile length, or about 100 ft (30 m) for a 10 ft (3 m) missile. In actual practice the impact velocity and penetration depth must be well below this to ensure the contents are not severely damaged."[31] It is assumed that the explosive device can be successfully delivered kinetically to the center of a 200-m-diameter asteroid; anything larger may require the use of some sort of drilling or auger device.

Table 16. Qualitative considerations for threat mitigation using nuclear fragmentation.

First-Order Qualitative Considerations	
Susceptibility to dust cloud	Low
Ability to handle target rotation	High
Requires landing on target	Maybe
Usefulness on fragmented body	Low
Swarm option	Medium
Second-Order Qualitative Considerations	
Usefulness as weapon	High
Perceived safety	Low
Synergy with other NASA missions	
Manned missions	Low
Robotic missions	Low
Resource utilization missions	Low
Costs	
Development	High
Deployment	Medium

5.2 Nuclear Deflection

Nuclear deflection of an asteroid or comet would probably be accomplished by using a nuclear device to produce a highly intense radiation burst. This is done while the device is still a significant distance from the target object. A relatively thin layer of the body's surface absorbs the intense, high-energy electromagnetic radiation—mainly hard x rays or gamma rays—and is vaporized. The intense heating blows the surface layer off and the ejected mass imparts a reactive impulse to the body.

There are other ways in which a nuclear blast might be used to deflect an asteroid. Near-surface blasts will break up the surface layer and blow large pieces of physical debris into space. Once again, a reaction would be produced and act to deflect the body. Unfortunately, energy losses in the fracture process and the absence of a preferred direction of motion for the detached pieces both serve to limit performance.

Another option would be to place a quantity of some low atomic number material between the nuclear device and the asteroid. This material would then absorb the radiation and blast energy from the device, would vaporize, and the resulting debris would impact the asteroid. Although possibly effective if used on a solid asteroidal body, this technique would be less effective on a comet or dusty and/or soft asteroid.

A deep subsurface blast could also be used, but the nuclear device must somehow be buried or be designed to survive the high gravity loads of a deep-penetrating projectile. Unfortunately, such a technique would still dissipate a majority of the available energy in the fracture process. One could envision a situation where only a few large fragments are produced or where the pieces have very little separation velocity and would all still strike the Earth. Even worse, some of the pieces could draft behind others during atmospheric entry, causing as much or even more impact damage on the Earth's surface than one large piece.

The first process described above—and used in the following analysis—is similar to a laser ablation process. In fact, the energy deposition, absorption, and subsequent plasma expulsion are based on the same physical principles and governed by the same mathematical equations as for laser ablation. The asteroid deflection process for nuclear pulse/EPPP is accomplished with a more conventional nuclear device design and nominally requires two to ten separate devices to be successful. The primary reason for this is that nuclear energy is efficiently liberated during very high yield detonations. The energy deposited is then transferred to the in situ propellant—surface material from the target body itself—to produce a very efficient propulsive technique.

First-order qualitative considerations all received the highest score for this technology. Table 17 lists the rankings as all green. A dusty target body surface or dust clouds do not affect the radiation energy from the blast. The impulse may be slightly reduced, depending on the particular conditions, but much of the energy will still be deposited on the surface and will produce thrust. Even the dust cloud itself, should it absorb a great deal of the energy, will generate large pressures on one side of the body and so assist in the desired deflection. A tremendous advantage to such an intense flash of radiation is that target rotation is essentially inconsequential. Like a strobe light that seems to stop a moving object for an instant, the entire thrusting event is over in milliseconds and the momentum imparted to the object is in one direction. Precise targeting or landing on the body is not needed and simplifies the mission requirements considerably. Fragmented asteroids or comets could be dealt with very well, since each pulse would affect the trajectories of

Table 17. Qualitative considerations for threat mitigation using nuclear deflection.

First-Order Qualitative Considerations	
Susceptibility to dust cloud	Low
Ability to handle target rotation	High
Requires landing on target	No
Usefulness on fragmented body	High
Swarm option	High
Second-Order Qualitative Considerations	
Usefulness as weapon	High
Perceived safety	Low
Synergy with other NASA missions	
Manned missions	High
Robotic missions	Medium
Resource utilization missions	High
Costs	
Development	Medium
Deployment	Medium

all objects simultaneously. Depending on the circumstances, there are several advantages that could be capitalized upon so as to maximize the effectiveness of the nuclear pulses. Finally, the swam option is also every effective with this technology as, after each pulse, an assessment can be made as to the effectiveness and implications of the event before the next one is sent. One large device might be used to obtain the maximum efficiency of the fission fuel, but this would not be a critical factor to mission success or cost. Many plasma pulses will allow precise control, reduce risk, and be more flexible in target engagements.

The second-order qualitative considerations have a mixed assessment ranking. The technique's usefulness as a weapon is obviously high, but this point is a little misleading. The very fissile material that would be used in an asteroid deflection system currently resides in nuclear weapons around the world. Removing this material from these weapons, and instead, using it for planetary defense, must surely produce a net benefit. Preexistence of the fissile material is also one of the reasons why the cost is not high for this option. In fact, besides the expense for nuclear safeguards, which might be accounted as an expense already being carried by other government agencies, particularly if excess weapon stockpiles are used, this should be a low-cost option. Perceived safety is a clear problem. However, in reality, nuclear weapons of this caliber have been safely maintained for over 50 yr without incident. A space-based planetary defense system would be far more secure, and a ground-based system would be no more difficult or expensive to maintain than the present silo-based ICBM systems already maintained by several countries. Synergy with other NASA missions is high, as the physics and many of the specific technology challenges would be applicable to an EPPP; e.g., nuclear pulsed propulsion, vehicles capable of interplanetary travel with acceptable trip times for human missions, or for deep-space robotic missions. Of course, extraterrestrial resource utilization would be greatly facilitated if this technique were available to deliver asteroids or comets to strategic locations in the solar system.

There are two primary hardware components to the nuclear deflection option: the nuclear device and a small rocket delivery system. For this assessment, the estimate for the nuclear devices is only based upon the fission of U235. Tritium cores for fusion enhancement and plutonium fuel were not considered but would be significant performance enhancers. The use of only uranium fuel is helpful because it is essentially nonradioactive and can be easily handled and stored for long periods of time. The performance for this application is excellent and little is gained by going to hydrogen bomb performance. Although, it may be easier to utilize existing and well-proven devices rather than build new ones, even if the design used a successful early-generation uranium mechanism.

The calculational process begins using inputs for the asteroid mass and its average density. Some nuclear device characteristics are defined: nominal yield, collimation factor, burst half-angle, burnup fraction, and inert material fraction, along with the standard properties of U235 (185-MeV fission fragment energy and uncompressed density of 19,100 kg/m^3). The ideal potential energy in the fuel (J/kg) is found from the energy release in megaelectron volts multiplied by Avogadro's number and then divided by the molecular weight—provided the proper unit conversions are also used. From this value, the desired yield, and the burnup fraction, it is straightforward to determine the required mass of U235. The total device mass is then the sum of the fuel and inert mass, where the inert mass is based on the estimated inert mass fraction (F_{inert}) given in equation (42):

$$M_{inert} = \frac{M_{U235} \times (\frac{F_{inert}}{100})}{1 - (\frac{F_{inert}}{100})} \ . \tag{42}$$

It should be noted that this analysis only assumes the kinetic energy from a pure fission reaction of U235, which is converted to intense thermal electromagnetic energy in the hard x-ray or gamma-ray wavelengths. The analysis neglects the neutron energy and prompt gamma ray produced, which would improve performance. Finally, a simple pulse unit volume is determined by assuming that the inert material (average) density is simply one-fourth of the density of the fuel, with an additional 10-percent fluff factor on the entire device.

The planetary body size is determined by assuming a roughly spherical shape and using the mass and density to estimate a radius. The asteroid is considered to be a hemisphere, illuminated at the optimum pulse unit distance. The standoff distance—perpendicular distance from detonation point to body's surface—is selected so that, at the device's cone half-angle, one-half of the exposed surface is irradiated; i.e., one-eighth of the total spherical asteroid surface. The geometric relationships are illustrated in figure 42 and the standoff distance is calculated using equation (43). The average energy deposited on the surface is based on the volumetric ratio of the radiation cone and the total spherical volume surrounding the detonation (eq. (44)), with a conversion factor of 4.186×10^{12} to convert energy from kilotons to Joules, and divided by the surface area (πlR^2) on which the energy is deposited. To account for the losses that might be involved; i.e., surface reflection, scattering, dust particle interference, reradiation losses to deep space, etc., only 25 percent of the energy just determined was regarded as available to provide thrust:

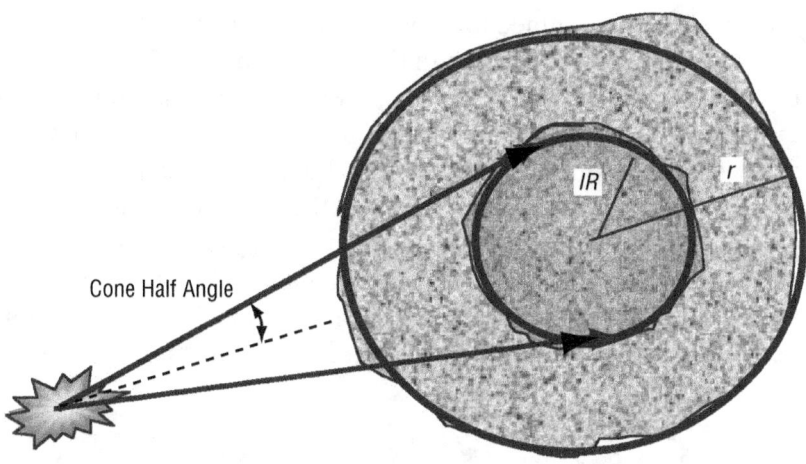

Figure 42. Geometric position of the pulse unit to the planetary body and cone half-angle definition.

$$D_{standoff} = \frac{IR}{\tan(\alpha)}, \text{ where } IR = \sqrt{r^2/2} \qquad (43)$$

$$E_{avg} = \frac{F_{collimation} E_{yield} \left(1/3 \pi D_{standoff} IR^3\right)}{\left(4/3 \pi D^3_{standoff}\right)\left(\pi IR^2\right)} . \qquad (44)$$

The depth of energy penetration was estimated at 20 cm, and this volume of surface material was used to estimate the in situ propellant available. The average velocity of the propellant leaving the body after one nuclear radiation pulse is calculated from the kinetic energy equation ($E = 1/2\ mV^2$) using the mass of the propellant (derived from the disk volume—for the 20-cm-deep region—multiplied by the average density assumed for the planetoid) and the energy as determined above. Not all of this material will be ejected in such a way as to contribute momentum to the planetoid in the desired direction. Therefore, a further efficiency factor of 20 percent was applied before determining the effective I_{sp} of the technique.

The calculation continues with the deflection ΔV for the body being used to calculate the total propellant needed at the I_{sp} predicted for the pulse unit interaction. The standard exponential rocket equation is used (eq. (45)), and the total number of pulses required to accomplish the deflection is found by dividing this result by the propellant mass previously calculated for a single nuclear device. Note the value is rounded up to the next whole number and no contingency nuclear pulse units were assumed to be carried in this initial analysis:

$$M_f = M_i e^{-\frac{\Delta V}{g I_{sp}}} . \qquad (45)$$

A rocket delivery system was assumed to provide standoff distance from the spacecraft (assumed to be an ORION or Gabriel type derivative as shown in fig. 34) and to provide some placement control of the fission device. Figure 43 illustrates the operation of the EPPP spacecraft and rocket delivery system. The calculations for this component of the system were nonrigorous and merely serve to estimate the mass and volume that would be needed. A 1-km/s ΔV was assumed and the rocket equation used to estimate the rocket propellant required to carry the nuclear pulse unit to the asteroid. Table 18 shows the basic assumptions made to estimate the rocket size. The chemical propulsion system mass includes the tank and other subsystems that scale with propellant load. (It is recognized that this is not the most accurate methodology, but within the framework of this study, it is an acceptable expedient.) The tank volume was estimated by assuming all the propellant was hydrogen with a density of 71 kg/m^3 and by specifying a 2-m-diameter cylinder tank.

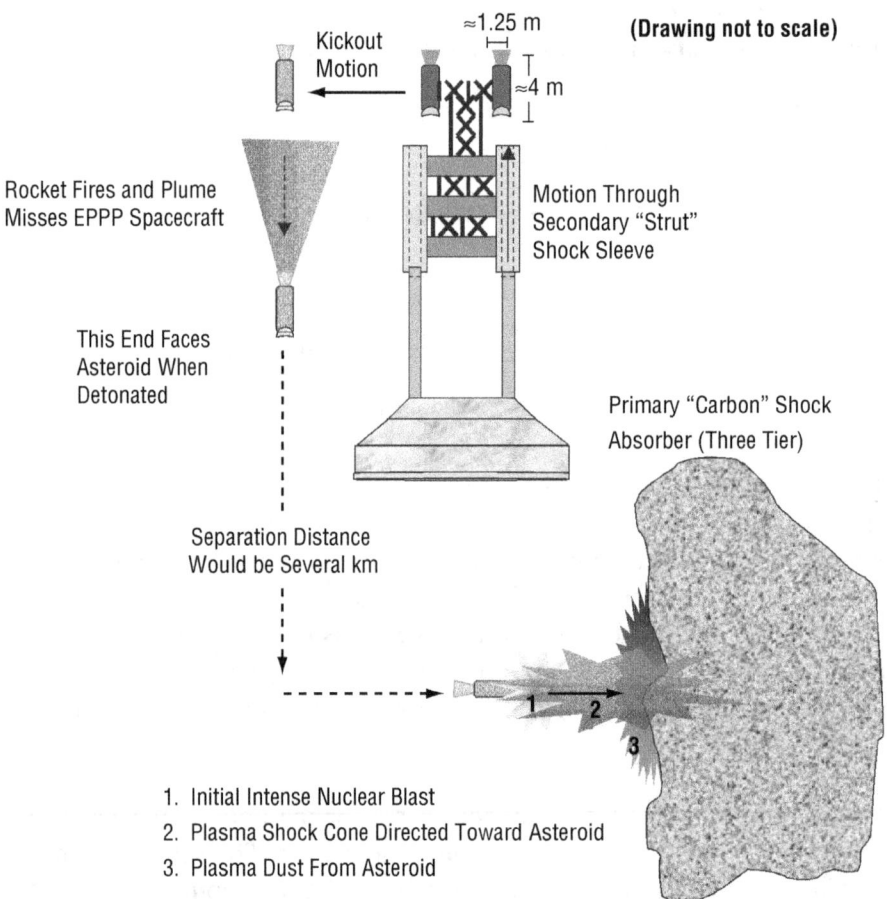

Figure 43. Nuclear pulse rocket delivery system sketch.

Table 18. Chemical rocket assumptions.

Characteristic	Value	Units
Specific impulse	450	s
Chemical propulsion system mass	500	kg
Nuclear device mass structure contingency	15	Percent
Integration structure and contingency factor	30	Percent
Pulse unit volume contingency factor	30	Percent
Nuclear device rocket volume contingency	15	Percent
Volume integration structure and contingency factor	20	Percent

The final part of the analysis confirmed that the fission pulsed detonation methodology is very insensitive to many of the input parameters and can easily move most threatening bodies safely away from the Earth. The system's qualitative parameters ranked very high, and even the secondary considerations that were unfavorable appear to have some mitigating arguments. The uranium fission technology assumed is well established, perhaps even to the point of being outdated, and the general estimates prepared were all conservative in nature. The plasma interaction with the planetoid body surface was not modeled, but expected propulsive results were established with significant efficiency degradation factors taken into account. The need for a second rocket delivery system with an arbitrary 1-km/s ΔV seems prudent for protection of the carrier spacecraft. Using an all-chemical system to deliver the nuclear devices from the Earth seems to be impracticable because the intercept time would be exceedingly long, although this trade space was not explored in detail.

5.3 Solar Sails

Solar sails are found in the literature as nonnuclear options for deflection. Sails were initially considered as possible deflection mechanisms. Using calculations shown in section 4, the overall performance of a solar sail can be determined where an asteroid or comet has somehow been connected to it and thus constitutes part of the payload.

An optimistic calculation of the overall ΔV imparted to the vehicle plus object is determined by using

$$\Delta V = a_0 \times t \; , \tag{46}$$

where a_0 is the characteristic acceleration, defined in the outbound solar sail description, and t is the time during which the sail is in operation. ΔV versus time can then be plotted for several sail areal densities and sizes. Figure 44 shows the plot of an object that is assumed to be a 10-m-diameter carbonaceous chondrite asteroid. It is obvious from the figure that the sail areal density is insignificant for this application. The payload mass of the asteroid dominates the sail mass. The lines indicate sail areas of 10^4, 10^5, 10^6, and 10^7 m^2, respectively. These areas translate to side lengths for a square sail of 100, 316, 1,000, and 3,162 m, respectively. It is expected that the deflection requirements for an incoming asteroid with <1 yr to deflect will be around 1–10 m/s. Therefore, all four size sails are able to deflect a 10-m object.

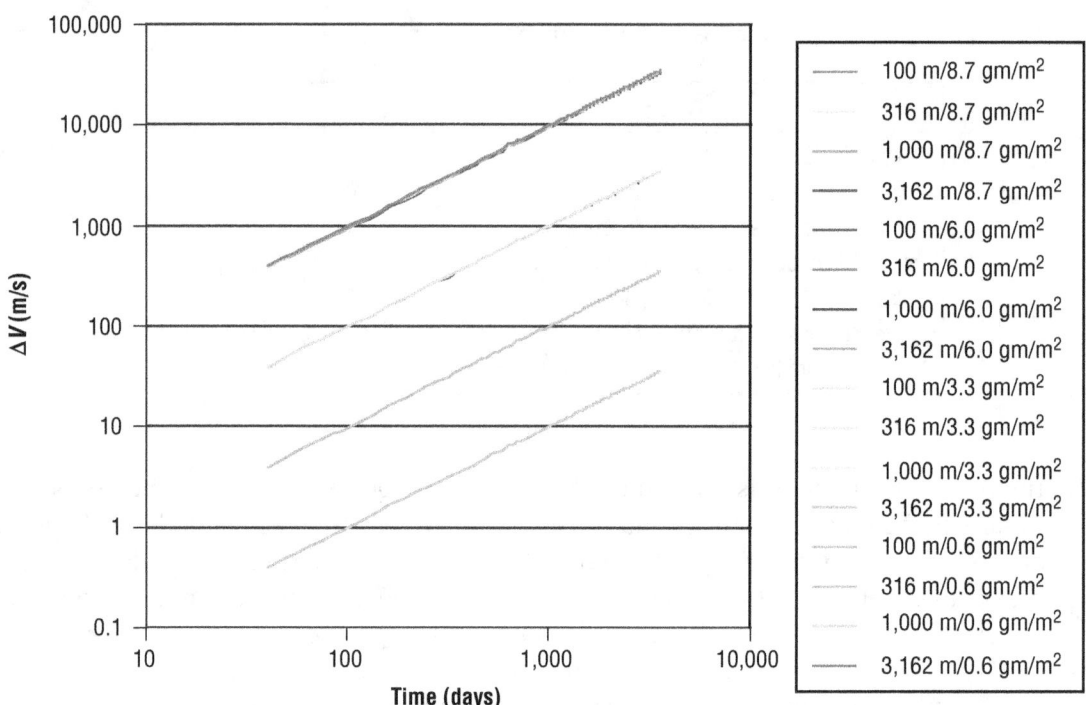

Figure 44. Deflection ΔV imposed on 10-m-diameter asteroid.

Of course, a 10-m object is of little concern. From the above, ΔV versus time can be plotted for asteroids of four different diameters: 10, 100, 1,000, and 10,000 m (fig. 45). From here, all three larger sails are able to handle deflection requirements for a 100-m object without an undue amount of time (1 yr). The largest sail might be able to deflect a 1-km object but would require 5 yr or more. Finally, deflection of a 10-km object seems outside the capabilities for any reasonably sized sail. From these results, using this option will necessitate the use of very large sails and will still not provide the capability of deflecting the largest asteroids and comets.

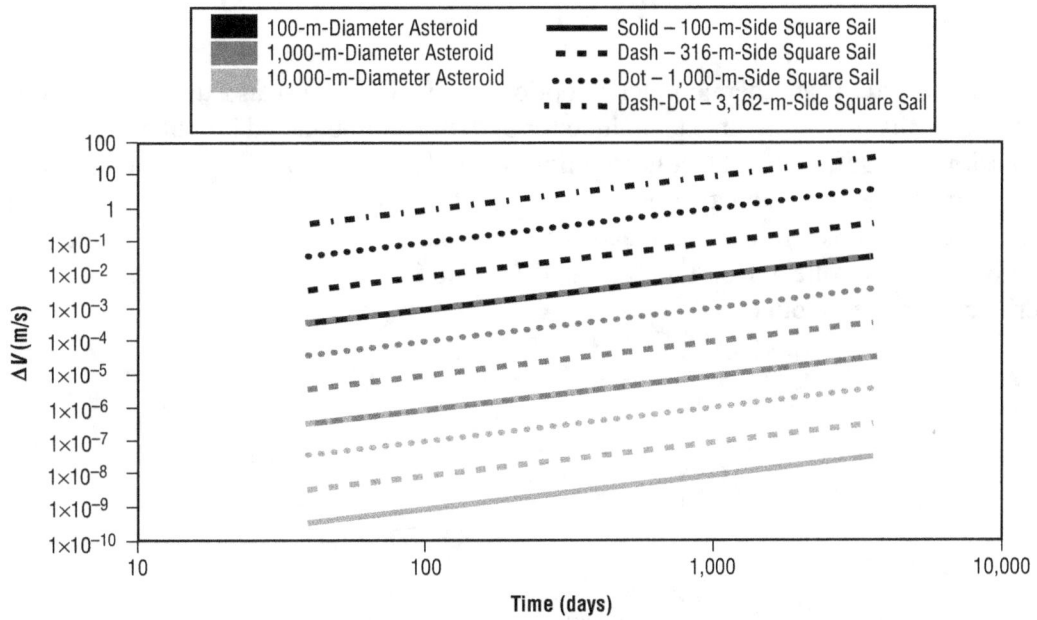

Figure 45. Deflection ΔV imposed by 6 gm/m² solar sail with varying areas on varying diameter asteroids.

Unfortunately, there are several complications that further exacerbate the sail's inability to deflect the most threatening objects. As indicated in table 19, any clouds of dust around the object would degrade the sail material. Most concepts envision the sail somehow encircling the object with a line or net. The force from the sail would then be transmitted through the line to the object. Obviously, this idea would have great difficulty if the object were rotating. The sail could be "rolled up" onto the asteroid or comet. For similar reasons, the sail would not be able to affect more than one fragment of a comet at a time. Finally, the sizes indicated for these objects are such that it is difficult to see how more than one sail could operate at a time.

Table 19. Qualitative considerations for threat mitigation using a solar sail.

First-Order Qualitative Considerations	
Susceptibility to dust cloud	Medium
Ability to handle target rotation	Low
Requires landing on target	Yes
Usefulness on fragmented body	Low
Swarm option	Low
Second-Order Qualitative Considerations	
Usefulness as weapon	Low
Perceived safety	High
Synergy with other NASA missions	
Manned missions	Low
Robotic missions	High
Resource utilization missions	High
Costs	
Development	Medium
Deployment	Low

5.4 Solar Collector

The solar collector is another nonnuclear option considered for use as a threat mitigation technique. It also offers synergy by using the same technique for both the outbound and inbound legs of a rendezvous-deflection mission. For deflection, the collected light from the curved sail is directed onto the asteroid by the secondary mirror, shown schematically in figure 46. It is noted that this concept excludes the propulsion system-needed stationkeeping. The added weight of such a system would substantially degrade the performance of the solar collector on the outbound leg and is most probably superfluous in view of the forces experienced on the inbound leg.

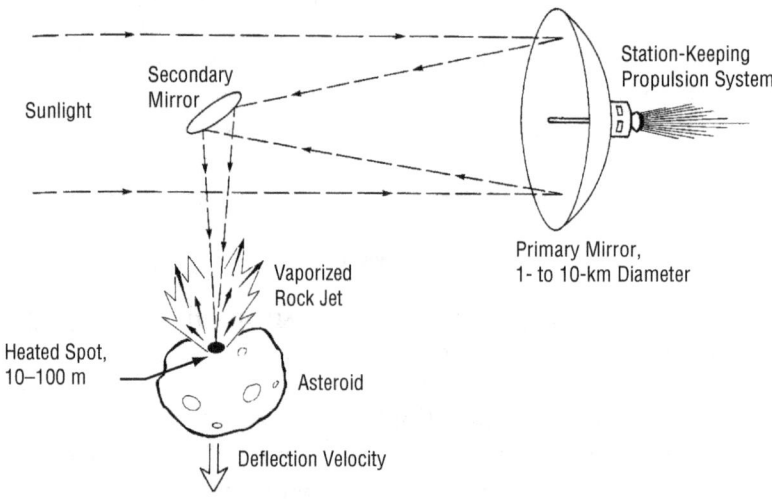

Figure 46. Schematic of solar collector.

The beam intensity incident on the object is assumed to be equal to that incident on the collector. Therefore, no losses are included for absorption or the effect of a noncolumninated beam. This is a little optimistic, but losses must be strongly avoided or the design temperatures required for the collector and mirror will become untenable. The beam intensity incident on the object is then given by

$$I(r) = P(r) \times c \ . \tag{47}$$

The pressure as a function of radius was defined in section 5.3 and c is simply the speed of light in a vacuum. In all respects, the analysis for the collector is the same as in section 5.3. All that is required is to determine the force imposed on the asteroid or comet by the incident beam. The following discussion deals with this.[32]

The amount of mass ejected from the object is calculated as

$$\dot{m}_e = \frac{I(r) A_{sail}}{H_{vap} + \frac{v_e^2}{2}} \ , \tag{48}$$

where H_{vap} is the heat of vaporization (3 MJ/kg for water ice) and v_e is the velocity of the ejecta. As a first approximation, the ejecta is assumed to come out at its sonic velocity (≈1 km/s). From this, the force exerted on the object by the ejecta can be determined:

$$F = \beta' \dot{m}_e v_e ,\qquad(49)$$

where β' is a gas expansion factor—assumed to be 0.5—which represents a hemispherical expansion.

For the purpose of this study, the orbital radius of the vehicle from the Sun was assumed to be 1.5 au. This is an assumption based on the average distance during the time period when the collector is acting on the object. Additionally, 500 kg of payload was assumed to account for avionics and other systems. The power of the incident beam at 1.5 au is nearly 5 GW for a 100-m-diameter collector. Obviously, a secondary mirror reflecting this much power will be a major design challenge.

A parametric of ΔV imposed on an asteroid or comet similar to that for the solar sail above can be created (fig. 47). Here, a solar collector with a diameter of 100 m is acting on asteroids with diameters of 10, 100, 1,000, and 10,000 m, respectively. The collector can obviously handle asteroids of up to 100 m without a problem. The intensity of the beam focused on the asteroid will scale with the square of the diameter of the solar sail. Thus, there is the expectation that a 1-km solar collector may be able to deflect asteroids of 1-km diameter. There is no obvious path for solar collectors to deflect asteroids of larger diameter.

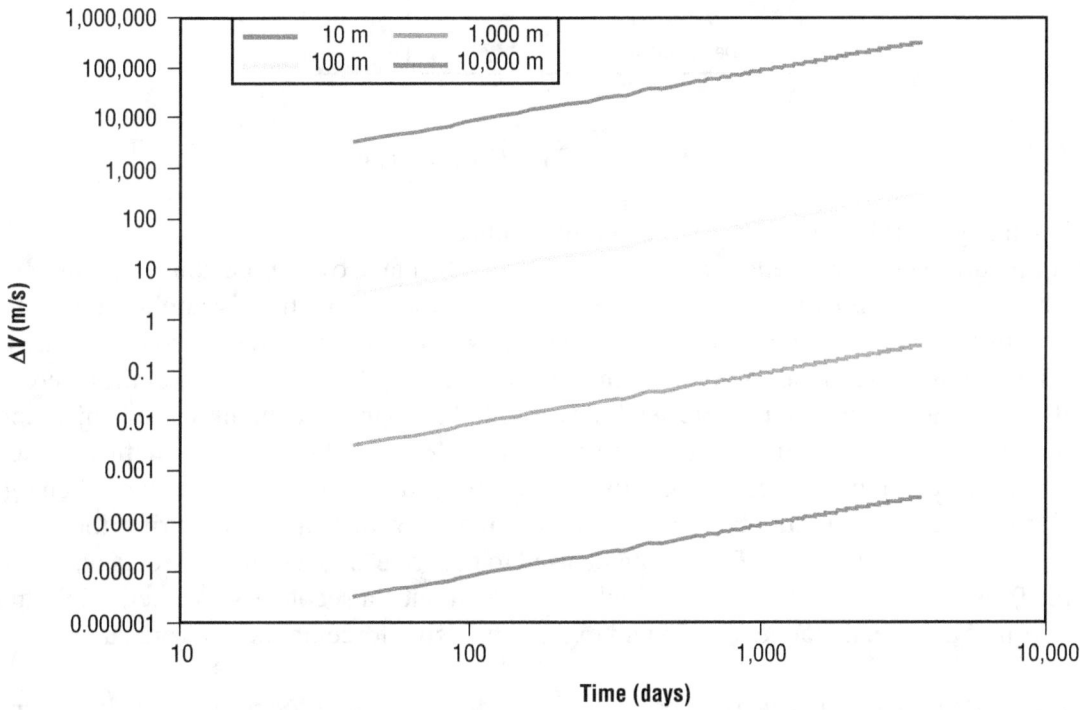

Figure 47. ΔV imposed on varying asteroid sizes by 100-m-diameter solar collector.

Any dust or ice clouds surrounding the object will degrade the beam from the solar collector. It will be difficult to match rotation with a rotating object. In this case, the incident beam will strike not a single site, but instead, a band around the rotating object. The need to constantly heat new portions of the object— or at least portions not heated since the previous revolution—will most likely result in a lower temperature and therefore slower ejecta. This will result in less thrust than could be achieved on a nonrotation body. For a fragmented body, although the beam can be directed on different components, only one can be pushed at a time. Finally, due to the size of the solar collector, it would be difficult for more than one to operate on an incoming object at a time. See table 20 for solar collector qualitative considerations.

Table 20. Qualitative considerations for threat mitigation using a solar collector.

First-Order Qualitative Considerations	
Mitigation by dust cloud	Medium
Ability to handle target rotation	Medium
Requires landing on target	No
Usefulness on fragmented body	Medium
Swarm option	Low
Second-Order Qualitative Considerations	
Usefulness as weapon	Low
Perceived safety	High
Synergy with other NASA missions	
Manned missions	Low
Robotic missions	High
Resource utilization missions	High
Costs	
Development	Medium
Deployment	Low

5.5 Magnetic Flux Compression

Of the many asteroid deflection approaches researched for this TP, none used electromagnetic forces. By their very nature, electromagnetic forces could offer an advantage over more conventional explosive or kinetic impact forces. With explosives and kinetic impacts, the momentum is applied through either a pressure force from the highly random kinetic thermal energy in a gas or at a localized impact area. Both of these approaches loose a significant amount of energy by increasing the temperature of the asteroid surface and possibly breaking the asteroid into pieces. Many researchers agree that breaking the object into pieces could make the problem more difficult and would be an undesirable effect of any deflection technique, except in the case of complete pulverization. With electromagnetic forces, a large amount of energy can be converted into a directed Lorentz force without producing a lot of heat or fracturing the asteroid into pieces. A rapidly changing magnetic field could be used to generate large countercurrents on the surface of the asteroid. These surface countercurrents would in turn produce a secondary magnetic field that would repel away from the original magnetic field and exert a repulsive force on the asteroid surface.

Any conductor in the presence of a changing magnetic field will produce an electric current proportional to the magnitude of the magnetic field and its rate of change. This current is due to the induced electric field produced in the space that the conductor occupies, and is defined by Lenz's Law:

$$\varepsilon = d\phi_B/dt \; , \tag{50}$$

where ε is the induced electromagnetic field, ϕ_B is the magnetic flux, and t is time. Knowing the magnitude of the electromagnetic field and the electrical resistance of the conductor (R), this resulting current can be calculated using Ohm's Law:

$$I = \varepsilon/R \; . \tag{51}$$

Electrical resistance can be further defined by the resistivity parameter (ρ):

$$R = \rho L/A \; . \tag{52}$$

Resistivity has units of ohms/meter; L is the length of the conductor, and A is the cross-sectional area of the conducting path. Resistivity has a specific value for different materials and also changes with material temperature. The inverse of resistivity is known as conductivity (σ). Knowing the geometry and material of a certain conductor, one can calculate with some certainty the electrical resistance.

Magnetic flux is simply defined as the product of the magnetic flux density (B) and the surface area (A) that the flux lines pass through:

$$\phi_B = B \times A \; . \tag{53}$$

If a particle of charge (q) moves with velocity (v) in the presence of both a magnetic field (B) and an electric field (E), it will experience a force (F) defined by the Lorentz equation:

$$F = q(E + v \times B) \; . \tag{54}$$

If the electric field and magnetic field are perpendicular, the force will be applied in a direction perpendicular to both fields. For the case where charge is traveling through a conductor, the force is defined by the following equation:

$$F = V(J \times B) \; . \tag{55}$$

For a conductor with volume (V) and a current density (J) passing through it, both the flowing electrons and fixed ions in the conductor will experience a force (F) in the same direction due to the interaction of the electric field and the magnetic field (B). The magnitude of the force is dependent on the strength of the magnetic field, the rate at which electrical current flows, and the conductivity of the conductor; i.e., the poorer the conductor, the smaller the resulting force.

Since most asteroids are composed of carbonaceous chondrite materials, which are poor conductors, it may be necessary to use a target disk with a good conductivity value. This disk would mate up to the asteroid and serve as a pusher plate. Unlike a pusher plate used for pulsed nuclear or conventional explosives, this disk could be a relatively thin, lightweight metallic foil of sufficient area to be effective. The reason that this target disk can be much thinner and less massive than an explosion pusher plate is the difference between the methods by which the force is produced. With the pusher plate, the force is applied via the shock wave that is produced from the large amount of heat energy released from the explosives. Without the pusher plate, the explosives would ablate a surface layer off the asteroid and the pressure would be applied to a more localized area that may not be able to withstand the resulting stress. This could lead to cracks or pitting. The thick, strong pusher plate acts as a shield similar to a bulletproof vest. It has sufficient area to spread out the force and sufficient thickness to withstand the localized stress of the impact/explosion. With the target foil, the force is produced by a Lorentz reaction between currents and magnetic fields. This force is more evenly distributed through the entire target (like a sail) and the amount of heat flux released into the surface is greatly reduced. Energy is not wasted in vaporizing or demolishing the asteroid material. The thin foil could be efficiently packaged and only deployed once the asteroid is encountered. The target disk need only be thick enough to absorb the magnetic flux. Depending on the rate at which the flux changes, the disk could be very thin. The thickness necessary to absorb the flux is calculated using the following equation:

$$\delta = \sqrt{t_{pulse}/\pi\sigma\mu_0} \ . \tag{56}$$

Skin depth (δ) is a function of the pulse time (t_{pulse}), the conductivity (σ), and the permeability of free space (μ_0). As long as the foil thickness is greater than δ, the magnetic energy will be coupled completely into the foil. The faster the pulse, the thinner the foil can be.

Since most magnetic fields are curved, there will be a limit to the diameter of the target disk beyond which the field lines will reverse direction and produce forces acting in the wrong direction, as shown in figure 48. This would be very undesirable. The disk diameter is dependent on the distance from the magnetic field source to the target disk. This diameter must be optimized, since the critical disk diameter increases as the source moves farther from the asteroid (more surface area for the flux to be coupled to and easier to achieve) and decreases as the source gets closer toward the asteroid (stronger field strength, but may require more difficult maneuvering).

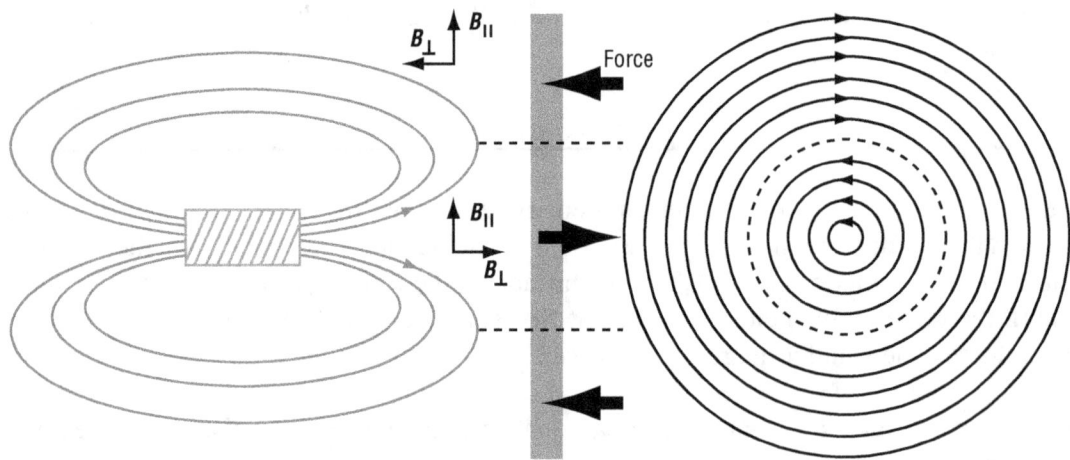

Figure 48. Critical disk diameter.

A common source of magnetic flux is a solenoid (coil) connected to a power supply. A solenoid produces axisymmetric, curved, nonuniform magnetic field lines with a similar field line profile to that of a bar magnet. The curved field lines may be divided into two component vectors—one normal to the target surface and one parallel to the target surface. The component normal to the target is responsible for inducing a current. The component parallel to the target surface is responsible for producing the Lorentz force. To accurately sum up the force produced by each varying field line, the Biot-Savart Law can be used:

$$dB = \frac{\mu_0 I}{4\pi} \frac{d\ell \times r}{r^2} . \qquad (57)$$

Using Biot-Savart, the magnetic field strength (flux density) (*B*) is calculated from knowledge of the coil path vector (ℓ) and the distance from the coil to the target (*r*). By summing up the length of the conductor path, the total field strength in a given region of space can be calculated. This magnetic field has energy proportional to the volume within the solenoid and the magnitude of the magnetic field:

$$E = V \frac{B^2}{2\mu_0} . \qquad (58)$$

Using equations (57) and (58), a parametric FORTRAN program was written to calculate the force that would be produced by an arbitrary solenoid design. If a 1,000-turn solenoid with a radius and length of 1 m, located at a standoff distance of 100 m, was supplied a current of 1 MA ramped up over a period of 10 ms, it would exert a force of over 9 million lb on a 100-m-radius target disk. That results in a very sizeable impulse. Although not impossible to produce, the logistics are not favorable. The inductance value of the solenoid in question is ≈3.95 H. This is roughly six orders of magnitude larger than most solenoids used in common electrical components. To force 390 kJ of magnetic energy into the solenoid in 10 ms would require over 4 billion V of electricity at a power of over 39 MW. This current and voltage level are not impossible to achieve, but they would require a very large power supply. Since anything more than one shot on the coil would be impractical (one would need a restoring force on the coil), it is not an economic solution. Current technology lightweight, high-power, disposable power supplies; i.e., batteries, capacitors, magnetic flux compression generators, etc., are incapable of providing this performance. Originally, this system was envisioned to use a magnetic flux compression generator.

A magnetic flux compression generator is an ingenious method for converting explosive energy into electrical energy by compressing a magnetic field. A solenoid containing a magnetic field has an energy associated with the volume of the solenoid and the magnetic field strength. From equation (58), it can be seen that the magnetic field strength plays a major role in the amount of energy within the solenoid. To increase the magnetic field, the current running through the coil must be increased. Another way of doing this is by compressing the field lines within the solenoid. By reducing the volume of the solenoid, the field lines are compressed and their strength is increased. Although the volume is decreased, since the energy is related to the square of the magnetic field strength, the energy is greatly increased. A convenient way of decreasing the volume is by placing a conducting sleeve filled with explosives along the centerline of the solenoid as shown in figure 49. This conducting sleeve is referred to as the armature of the flux compression generator and the surrounding coil is referred to as the stator. The stator

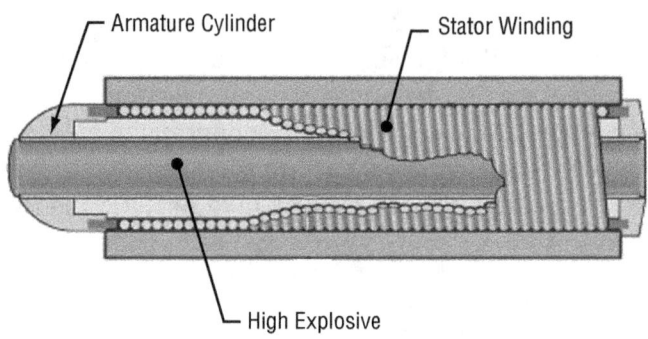

Figure 49. Magnetic flux compression generato.[33]

is energized with a current pulse, usually supplied by a capacitor bank, and a strong magnetic field is created within. An instant later, the explosives within the armature are detonated from one side. This explosion propagates along the armature, expanding the conducting sleeve as it goes. Because the armature sleeve is conducting as it moves through the magnetic flux within the solenoid, a countercurrent is induced. This countercurrent creates a magnetic field that repels the solenoid field and compresses it into the decreasing volume caused by the explosion. The magnetic field and resulting current at the end of the process are extremely large.

Using this method, currents in the mega-ampere range can be produced for a very brief period of time, usually on the order of nanoseconds. The flux compression generator can be coupled to a load or can be uncoupled to produce a very large electromagnetic pulse (EMP). The latter configuration is sometimes referred to as an E-bomb, since its EMP is often great enough to destroy nearby electrical equipment. The uncoupled configuration is shown in figure 50.

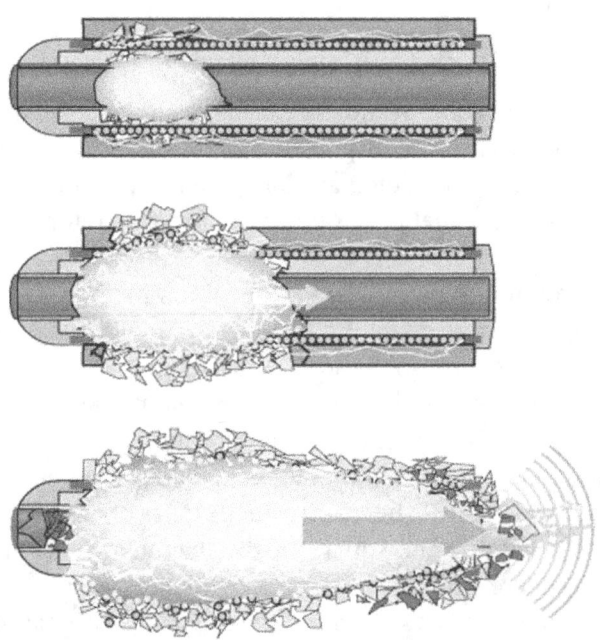

Figure 50. E-bomb magnetic flux compression generator.[33]

Unfortunately, magnetic flux compression generators provide peak powers in nanoseconds instead of milliseconds and their energy outputs are inadequate for driving the coil proposed for the asteroid deflection system. There is, however, another aspect of the magnetic flux compression generator that may have been overlooked at the beginning of this study and has not subsequently been explored. Although a flux compression generator would never be able to produce an EMP great enough to reproduce the performance of the proposed coil, another device may be able to do so.

Nuclear weapons detonated at high altitudes in the Earth's atmosphere produce a gigantic EMP. This was demonstrated by accident when a nuclear test conducted high in the atmosphere created an EMP that completely blacked out Hawaii and other parts of the United States. This EMP traveled down to the ground, disrupting and destroying electrical devices, batteries, generators, and wires. Since an EMP is simply a large traveling magnetic flux, it can couple with conductors in the same way in which radio waves couple with antennas. The major difference between the magnetic flux compression generator and a nuclear bomb is the way in which the EMP is produced.

When a nuclear weapon is exploded, a large amount of radiation in the form of gamma rays and x rays are released. These high-energy photons can be absorbed by elements with low atomic numbers, such as nitrogen and oxygen—two major components of the Earth's atmosphere. When these high-energy gamma photons strike an atom, they can knock loose electrons and send them flying away at great speed. This phenomenon is named the Compton Effect after its discoverer, A.H. Compton. These moving electrons serve as a large electrical current and produce a very large magnetic field for a brief period of time. This huge EMP can be coupled into conducting materials as electrical current. In the case of high-altitude nuclear tests, the atmosphere acts as a transformer, converting electromagnetic radiation (gamma rays) into lower frequency electromagnetic waves (radio waves). The amount of energy is conserved—if one ignores loss mechanisms—but its form is changed.

Perhaps a device could be constructed to use the compact and abundant energy released during a nuclear explosion to create an EMP suitable to drive currents in the foil target disk described earlier. To the author's knowledge, this idea has not yet been explored.

Except for the nuclear configuration—not yet explored—the magnetic flux driver concept is relatively safe and simple. It could be constructed from existing technology rather inexpensively. This concept could be very effective on fragmented and rotating bodies that have a high content of ferromagnetic materials, or an existing magnetic field. Nonmagnetic bodies would require use of a target disk, which would be less effective on rotating and fragmented bodies. Dust should not be a problem, since it would not block electromagnetic radio waves. The nature of this option would almost mandate that a swarm approach be adopted. See table 21 for quantitative considerations.

5.6 Mass Driver

The term mass driver is used to describe a variety of electromagnetic acceleration systems; all use electrical power to accelerate a bucket containing an inert mass to very high speed. The bucket is designed either without a top cover or with one that can be opened in flight, so that as it decelerates, the contents leaves the bucket and continues on at high speed.

Table 21. Qualitative considerations for threat mitigation using magnetic flux compression.

First-Order Qualitative Considerations	
Sensitivity to dust cloud	Low
Ability to handle target rotation	Low
Requires landing on target	Impact
Usefulness on fragmented body	Medium
Swarm option	High
Second-Order Qualitative Considerations	
Usefulness as weapon	Low
Perceived safety	High
Synergy with other NASA missions	
Manned missions	Low
Robotic missions	Low
Resource utilization missions	High
Costs	
Development	Medium
Deployment	Low

Although there are several different types of mass drivers,[34] all make use of magnetic attraction and repulsion between the moving (bucket) and the stationary parts of the device. A distinct type is distinguished by the specific shape and location of the magnets.

Although initially proposed[34] for space launch applications, mass drivers are also applicable to planetary body maneuvering. A special deployment vessel would transport the entire mass driver system out to the asteroid and install it on the surface. The system consists of the mass driver itself along with its electrical power source, either a nuclear reactor or a solar array, and its thermal control system. The basic rationale behind the use of a mass driver in this application is that it would make use of the asteroid surface material as reaction mass; i.e., expellant. As this material is already present, it is essentially free propellant, so even a system with a modest exhaust velocity; i.e., expellant speed, could offer good overall performance.

Two options are available regarding the provision of reaction mass:

(1) The deployment team could mine the entire stock of reaction mass before departure and place it into a suitable storage and dispensing system from which it could be fed to the mass driver as needed.

(2) An automated mining system could be deployed on the asteroid to operate continuously, together with the mass driver, after the deployment team has departed.

Option (2) is clearly more complex and raises additional questions over reliability and maintenance.

The intent of this study was to establish the general utility of mass driver systems for planetary maneuvering, not to determine which specific design offers the best performance, although this might be a suitable subject for a follow-on study. Accordingly, a relatively straightforward design was chosen to represent the entire spectrum of designs. The principal concern was to select a design that was relatively straightforward to model. This turned out to be the design with the greatest geometrical simplicity and symmetry—the coil gun.

The coil gun uses a series of equally spaced stationary solenoidal electromagnets (drive coils) to accelerate a bucket carrying a number of smaller but identically spaced solenoids (bucket coils). The concept is illustrated in figure 51.

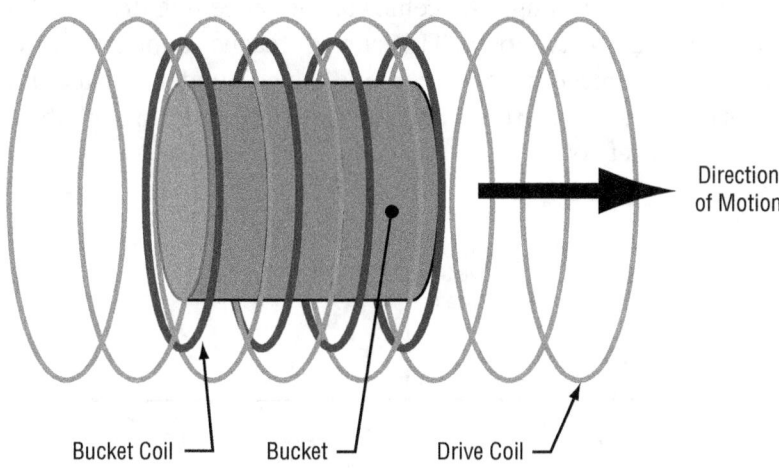

Figure 51. Coil gun conceptual design.

In the case shown above, which is that modeled, the bucket is surrounded by and attached to four current-carrying superconducting coils and is accelerated along the axis of the drive coils. Current flowing in each drive coil produces a magnetic field, a component that exerts an axially directed motive force on the bucket coils. The relative current flow directions of an arbitrary bucket coil and the nearest pair of drive coils is shown in figure 52.

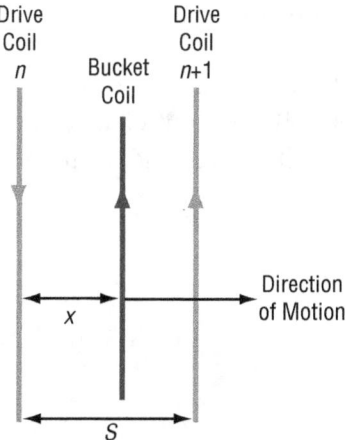

Figure 52. Relative current flow directions of bucket and immediately adjacent drive coils.

Coils with currents flowing in the same sense attract; those flowing in opposite senses repel. Thus, in the above example, the bucket coil is being repelled by the drive coil, which it has just passed (number *n*) and is being attracted to the next one (*n*+1).

Clearly, for this concept to work, each drive coil must reverse its current flow direction as a bucket coil passes through it, thus changing an attractive force for an approaching coil into a repulsive force for a receding coil. The current profile for one drive coil is shown in figure 53. The drive coil current is shown as a function of the bucket coil position (*S* being the distance between adjacent drive coils). When the bucket coil is at $-S$; i.e., just passing through the center of the preceding drive coil, the drive coil current begins to increase, producing an attractive force. This current reaches a maximum when the bucket coil is at $-S/2$; i.e., halfway between the preceding drive coil and the coil under consideration, and subsequently declines. It reaches zero as the bucket coil arrives and subsequently goes through the negative portion of its cycle, repelling the receding bucket coil.

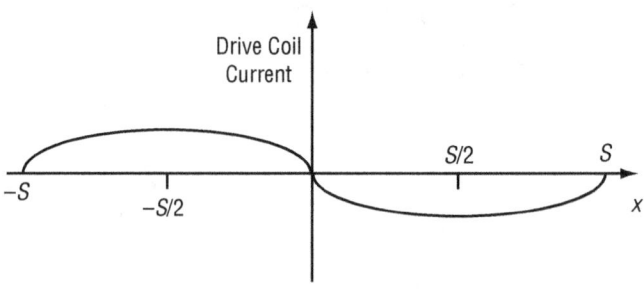

Figure 53. Relative current flow directions of bucket and nearest drive coils.

The design proposed here has four adjacent bucket coils separated by a distance *S*; i.e., identical to the spacing between adjacent drive coils. This means that, as bucket coil No. 1 passes through a drive coil, bucket coil No. 2 passes through the preceding drive coil. This, taken together with the drive coil current profile shown above, dictates that successive bucket coils must have their current flowing in opposite directions; i.e., exactly half a cycle out of phase.

The current directions for both drive and bucket coils at any instant is illustrated in figure 54. Note that the current flowing in each bucket coil does not change its flow direction, although its magnitude varies—due to induction—during the transit between adjacent drive coils.

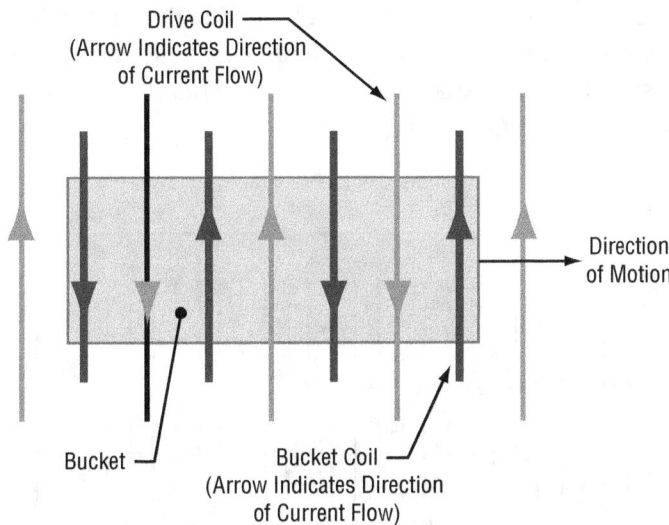

Figure 54. Relative current flow directions of bucket and surrounding drive coils.

To minimize unnecessary energy losses, it is anticipated that only those drive coils that immediately surround the bucket, together with a small number ahead and behind, will be energized at any instant. A minimum of five drive coils must be energized in order to cover the bucket itself. This can be seen clearly in figure 54. In reality, the number energized will depend upon the speed with which the bucket is moving and limitations imposed by current-switching technology.

It should be noted that although a bucket coil receives an accelerating force from the closest pair of drive coils, it experiences alternately retarding and attractive forces from each more distant pair. Fortunately, these more distant coils exert smaller forces than the nearest ones, illustrated in figure 55.

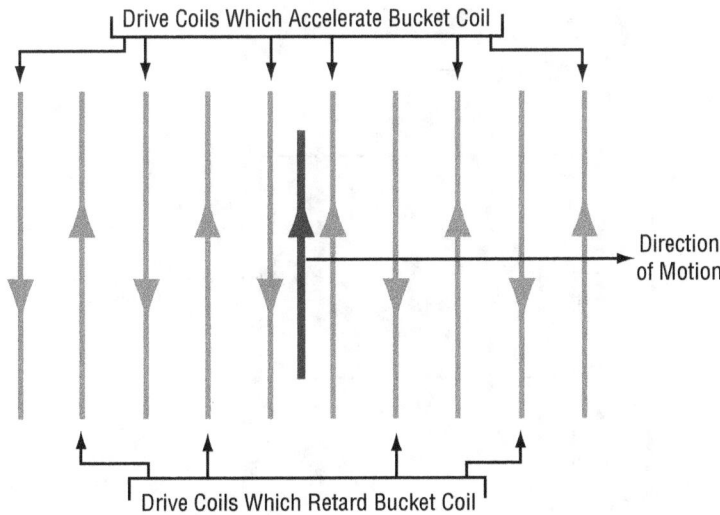

Figure 55. Relative current flow directions of bucket and nearby drive coils.

When the bucket and its contents have been accelerated to their maximum speed, the bucket is then decelerated by a set of braking coils. As the bucket decelerates, the nondecelerated expellant separates and continues at the maximum speed until it exits the mass driver. This is illustrated in figure 56, which shows three different times during an operational cycle.

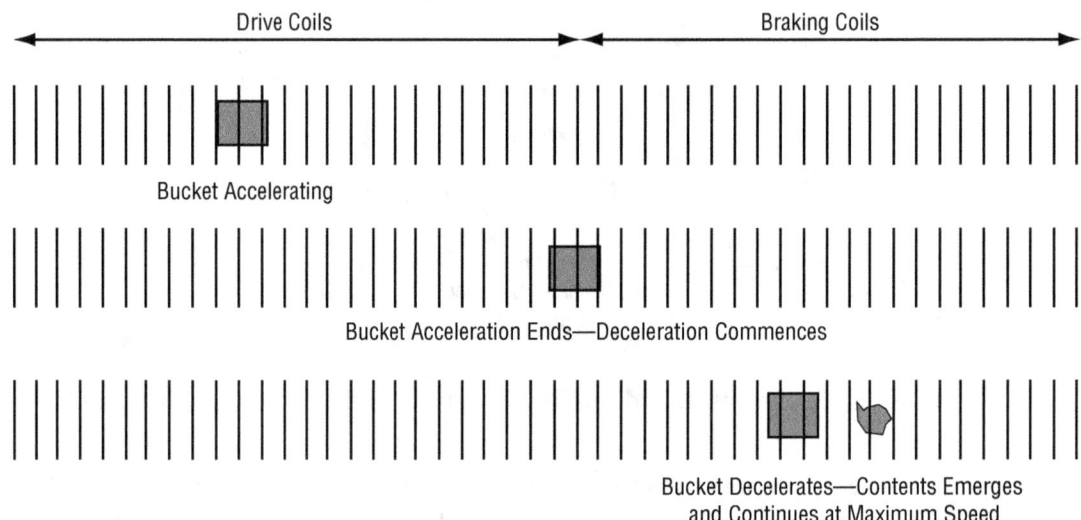

Figure 56. Mass driver operation.

The braking process is essentially the same as the acceleration process. The braking coils are of the same design as the drive coils. The coils are successively energized as the bucket coils pass them. The only difference is that during braking, the braking coils are energized exactly half a period out of phase with their drive equivalents. This means that the forces are in the reverse direction to those encountered during the drive section. Figure 57 shows the disposition of currents during braking. This should be compared with figure 54, the equivalent diagram for the drive coil portion of the mass driver.

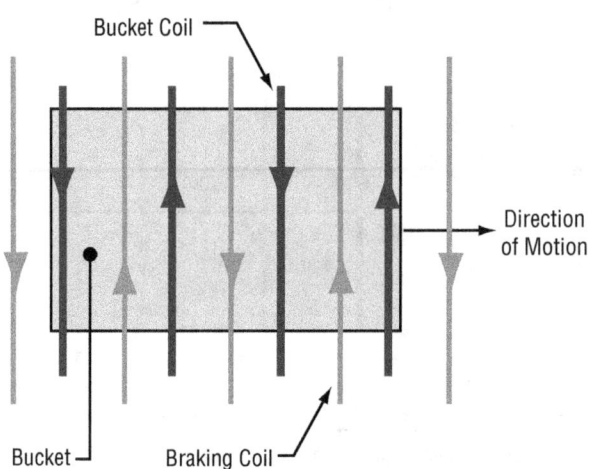

Figure 57. Relative current flow directions of bucket and braking coils.

In addition to providing a means of decelerating the bucket, the braking coils also provide a means of converting its kinetic energy back into electrical energy, thus making it available for use during the next cycle. Just as the drive coil circuits lose electrical energy as they accelerate the bucket, so the braking coil circuits gain electrical energy as they decelerate the bucket. A simple schematic of the entire mass driver system is given in figure 58.

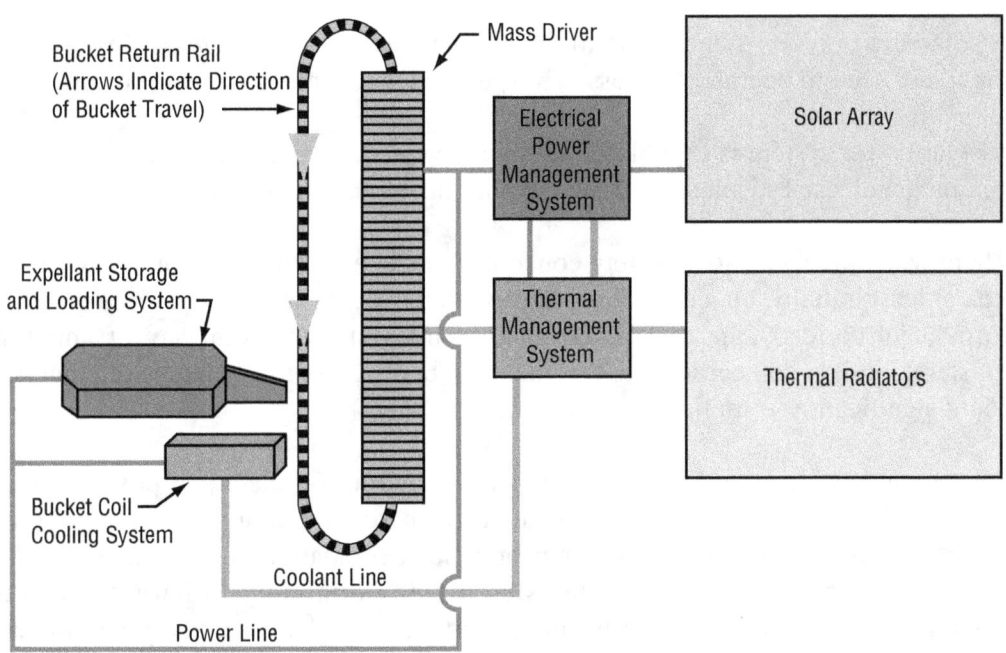

Figure 58. Mass driver system schematic.

The mass driver unit and return rail provide a complete closed circuit around which the bucket travels. After rapid acceleration and deceleration along the mass driver (coil gun) itself, the bucket moves at a relatively slow speed along the return rail. It halts at the expellant loading system in order to take on a fresh load of reaction mass and then proceeds on to the start of the mass driver.

While the loading process is taking place, the bucket is placed in contact with a cold plate whose function is to extract the waste heat generated within the bucket coils during the previous cycle. Although superconducting coils are used throughout the design, there is still some heat generated during the acceleration and deceleration processes that must be extracted to prevent the superconductor from rising to its critical temperature. The cooling system must also remove waste heat from the superconducting stationary coils, both drive and braking, between successive operational cycles. Connections between the thermal management system and both the mass driver and the bucket coil cooling system are shown above. An array of thermal radiators disposes of the waste heat.

There appear to be only two viable power source options currently available: a nuclear fission reactor and a solar array. For bodies sufficiently distant from the Sun, only the nuclear option is likely to be practical; however, for the purpose of this exercise, a solar array has been selected. This choice has been made for several reasons:

(1) Mass and performance modeling for a solar array are somewhat simpler than for a fission reactor, making it a more pragmatic choice regarding the analytical effort involved.

(2) A solar array power system currently poses fewer concerns over reliability, an important concern given the likely need to operate the mass driver, untended, over a long period of time.

(3) A solar power system is capable of operating even after sustaining damage. Parts of the array, which are nonfunctional, can be isolated and the system can continue to work, albeit at a lower power level.

(4) There would likely be some safety concerns associated with the placement of a fission reactor on a body; i.e., at least initially, on a collision course with the Earth. Obviously, if the mass driver system were to perform its job correctly, there would be no problems. But in the event, say, of a malfunction in the mass driver system, it might be necessary to fragment the body. If so, the presence of a nuclear reactor on its surface could significantly complicate the operation.

Against these considerations, it must be noted that the useful electrical power available per unit array surface area will be very low. Apart from being further from the Sun than is normal for solar power application, there will be additional causes of performance degradation. The array will almost certainly have to be simply laid on the surface of the asteroid and is unlikely to be normal to the incident radiation. If the asteroid is rotating, which is almost certain to be the case, additional losses will result. Finally, the process of preparing the expellant, regardless of whether it is done by continuous mining or by deployment vessel, will likely produce a large amount of dust and debris. The very low accelerations, both gravitational and of the body itself, will permit this dust and debris to remain in the vicinity of the mass driver for a significant time, allowing some of it to settle on the array, further reducing its effectiveness. Some details of the simple model used for the solar array are given in appendix B.

For the purpose of this study, it is assumed that expellant is prepared by the deployment vessel before its departure. The only other option—continuous automated mining—requires a detailed study in its own right. The problems of operating mechanically complex equipment, such as that needed to extract and render asteroidal material, in an ultra-low-gravity environment, are beyond the scope of the current work.

Although it is not the authors' intent to prejudge the results of this study, the following scenario is proposed for the application of mass drivers to planetary body maneuvering. A crewed deployment vessel conducts an extended mission in the vicinity of the asteroid belt. Its targets are a small number of asteroids whose orbits pose a threat to the Earth in a time period of, say, 10 to 50 yr into the future. At each of these asteroids, a mass driver is deployed and mining equipment, carried on the deployment vessel, is used to prepare a stock of reaction mass. At each asteroid, the mass driver commences operation and the deployment vessel departs for its next target. If necessary, subsequent visits can be arranged to conduct any needed maintenance and repair. Additional reaction mass can also be produced and stored during these visits.

The mass driver system deployed under this scenario is shown in figures 59 and 60. Figure 59 shows the entire system, which is dominated by the large solar array and thermal radiators. Figure 60 identifies the actual mass driver (coil gun) and associated equipment.

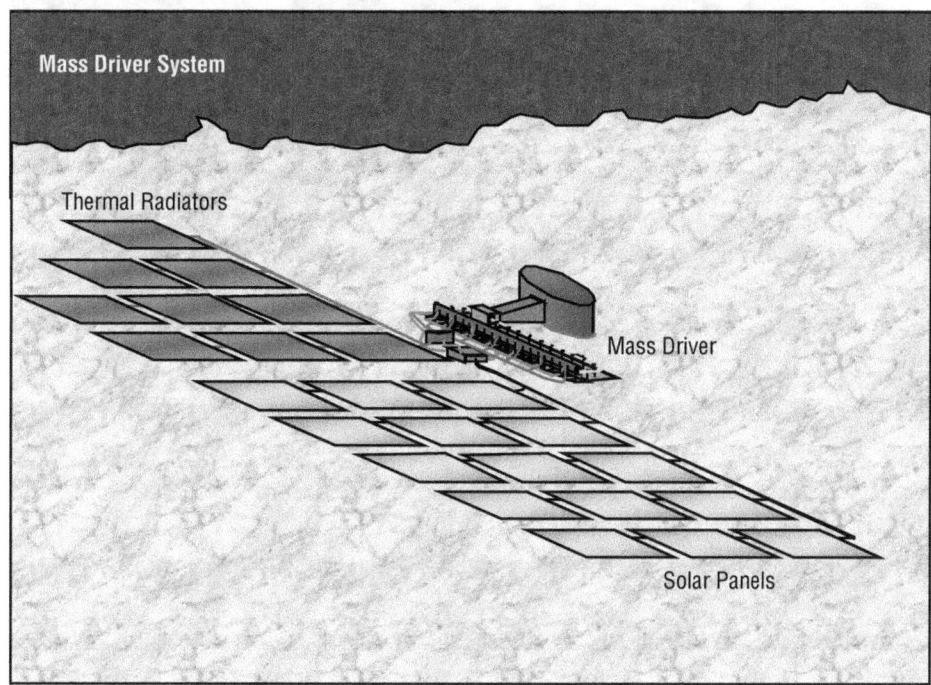

Figure 59. Mass driver system view.

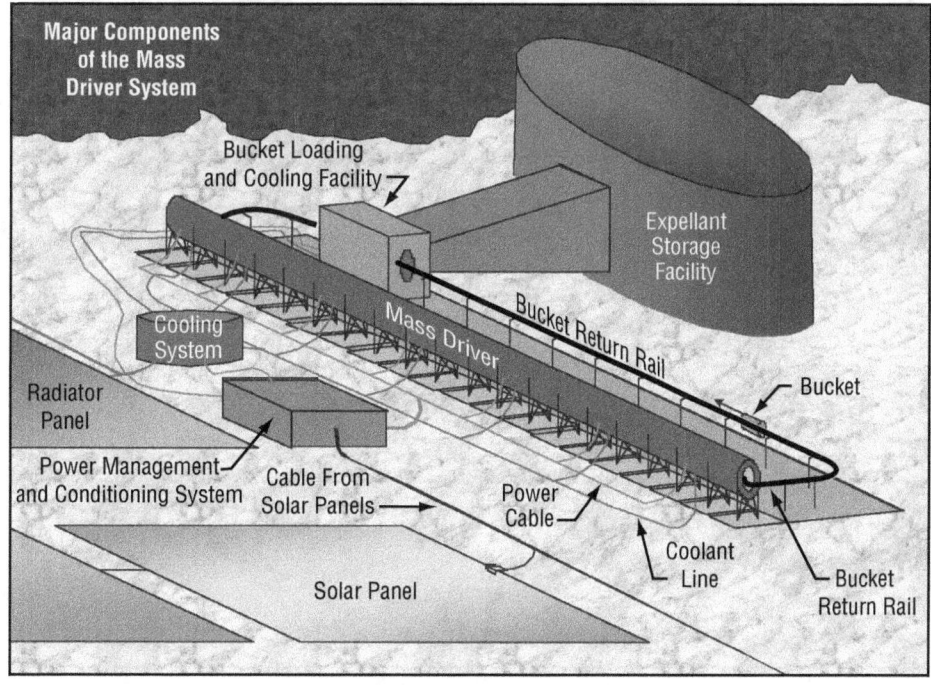

Figure 60. Main components of the mass driver.

Figures 61–66 illustrate system operations. Together, they show the system in operation through a complete cycle—the discharge of one bucket load of expellant.

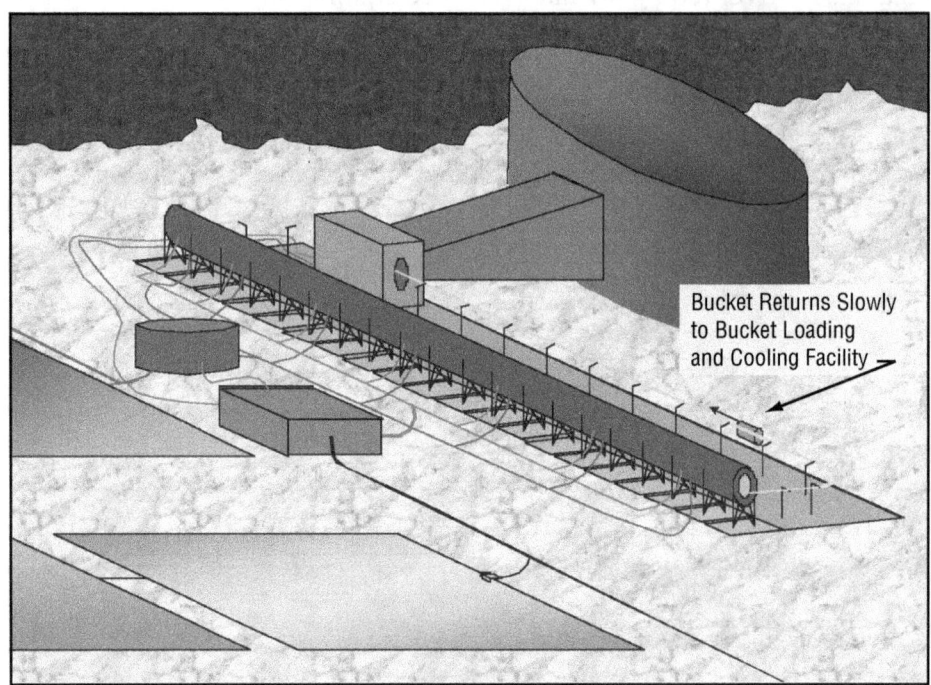

Figure 61. System following discharge—bucket returns for reloading.

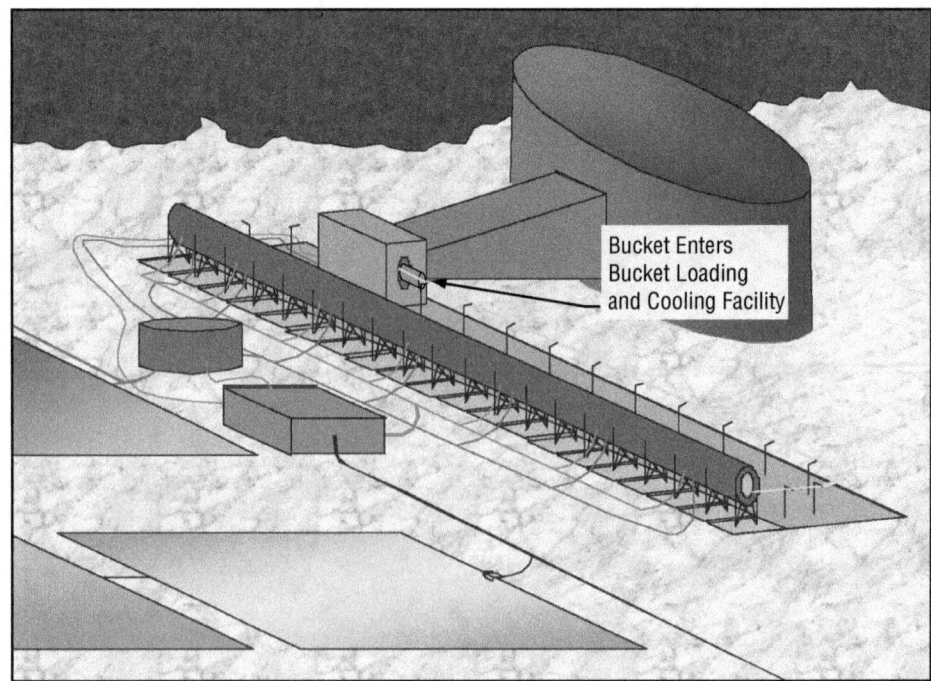

Figure 62. Bucket about to be reloaded and cooled prior to next discharge.

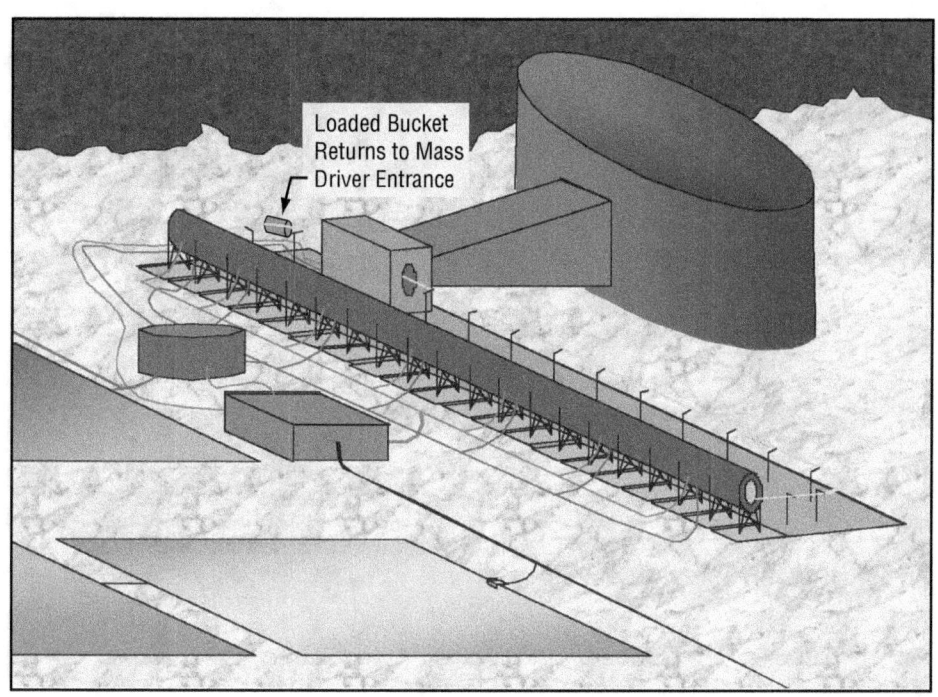

Figure 63. Loaded bucket about to enter mass driver.

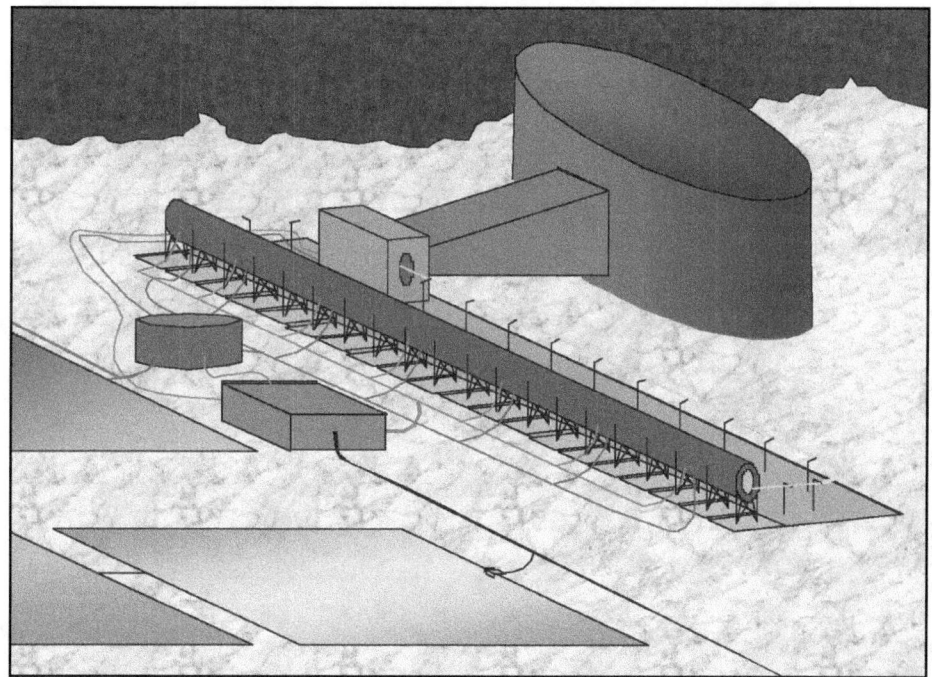

Figure 64. Bucket containing expellant under acceleration within mass driver.

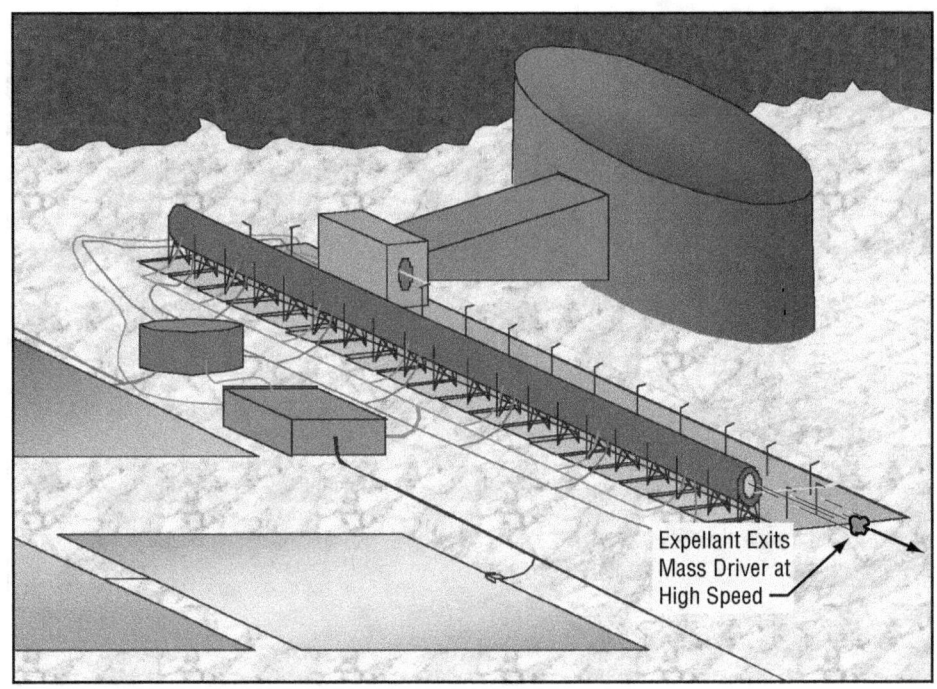

Figure 65. Discharge in progress—bucket is decelerating while expellant mass exits mass driver at high speed.

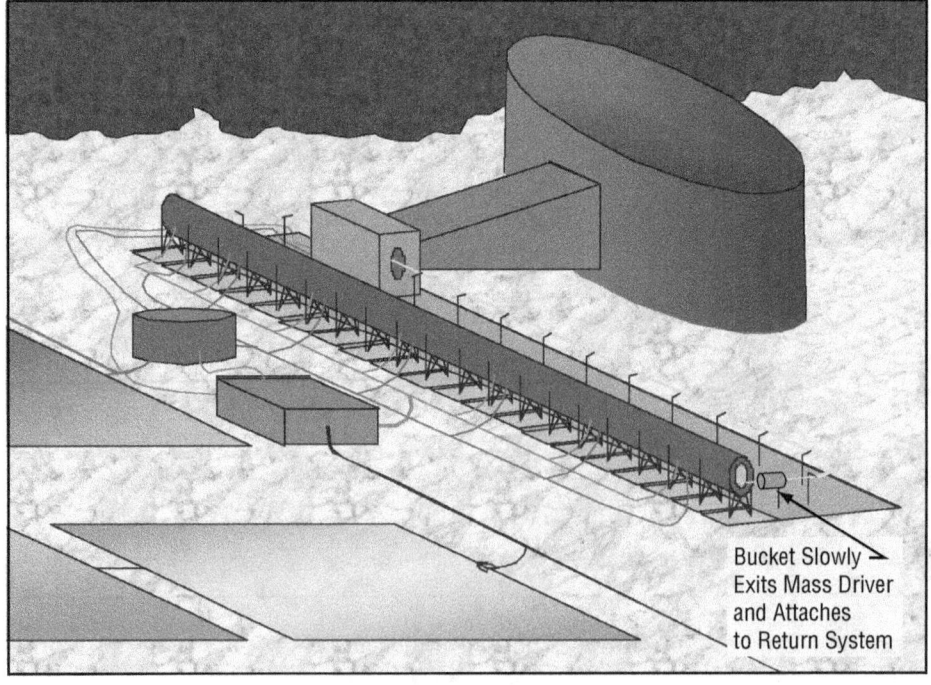

Figure 66. Following discharge—decelerated bucket exits mass driver and joins return system.

While there may be additional situations under which mass driver systems could be employed for planetary maneuvering, it is the above scenario that formed the basis of the analysis carried out during the present study. A simple performance and mass model for the mass driver and its principal supporting systems is presented in appendix C.

Table 22 lists the qualitative considerations that apply to the use of mass drivers for planetary body maneuvering. It is easier to conceive application of this technique to asteroid maneuvering than to comet maneuvering. A comet nucleus is likely to be both fragmented and dusty. A mass driver clearly cannot move more than a single body and will probably suffer some damage if operated in a dusty environment due to secondary impacts. Even if installed on a single asteroid in a nondusty environment, rotation will reduce the mass driver's efficiency. It will only be able to operate when oriented in the proper direction. To improve the technique's efficiency, two mass drivers could be deployed at appropriate "balanced" locations. These could first be used as necessary to counteract the asteroid rotation and then used together to provide a torque-free net force.

Table 22. Qualitative considerations for threat mitigation using the mass driver.

First-Order Qualitative Considerations	
Sensitivity to dust cloud	Medium
Ability to handle target rotation	Medium
Requires landing on target	Yes
Usefulness on fragmented body	Low
Swarm option	Medium
Second-Order Qualitative Considerations	
Usefulness as weapon	Medium
Perceived safety	High
Synergy with other NASA missions	
Manned missions	Low
Robotic missions	Low
Resource utilization missions	High
Costs	
Development	High
Deployment	High

The mass driver has a relatively high inherent safety rating against these potential complexities; this is largely due to our selection of a solar array to provide primary power. Use of a nuclear reactor, although in many ways more appropriate for the outer solar system, would reduce the safety rating. High-speed projectiles could cause significant damage, so a mass driver has some intrinsic weapons capability, but it would clearly not be a weapon of choice in any realistic situation.

5.7 Kinetic Deflection

Although potentially threatening planetary bodies will all be very massive—even a 30-m-diameter sphere of water ice would have a mass of ≈15,000 mT—in many cases, only a very slight orbital perturbation would be necessary to render them harmless. The energy required to carry out such perturbations, while still large, does not necessarily have to be supplied by an explosive event, whether conventional or nuclear. Even a relatively small interceptor craft, if accelerated to a sufficiently high velocity, possesses a formidable amount of kinetic energy that may be adequate to produce the desired deflection.

This raises the possibility of using a relatively small, inert interceptor body, raised to high speed, as a deflection tool. This technique offers the attraction that, once the interceptor's onboard propulsion system has exhausted its propellant, the remaining dry mass still fulfills an important function by virtue of the kinetic energy it carries.

There are clearly major issues relating to guidance of the interceptor, particularly during its terminal mission phase. With closing speeds perhaps in excess of 30 km/s, maneuvering capabilities will be limited and reaction times critical. For the purpose of this discussion, issues of guidance and terminal maneuvering will not be addressed. Clearly, some of the technologies being developed for terrestrial ballistic missile defense are likely to be applicable. Clearly, this subject is not appropriate for an open report.

This section contains the derivation of a simple model for use in evaluating the effectiveness of the kinetic deflection technique. The model is comprised of two distinct parts: (1) A largely analytical impact and momentum exchange model,[35] and (2) the issue of planetary body fragmentation.[36]

Figure 67 shows a general situation in which an interceptor is about to impact a planetary body. The velocity vectors of the body and the interceptor can always define a plane, so without loss of generality, the analysis can be presented in two dimensions. A modified inelastic model is given below. Although basically inelastic, allowance is made for the ejection of debris from the collision site.

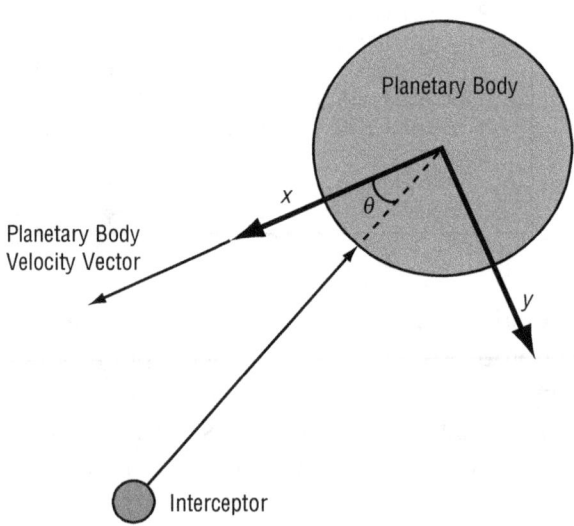

Figure 67. Interception geometry.

An *x-y* coordinate system is established as shown and then a momentum conservation equation is established as follows.

For motion along the *x* axis,

$$M_{NEO}u - M_i v_i \cos\theta = (M_{NEO} + M_i - M_{ej})(u - \Delta u_x) + P_{ej} \cos\theta , \qquad (59)$$

where

M_{NEO} = mass of the NEO before impact of interceptor
u = initial speed of NEO
M_i = mass of interceptor
v_i = initial speed of interceptor
θ = angle between NEO and interceptor velocity vectors (see fig. 67)
M_{ej} = mass of material ejected due to impact
Δu_x = change in *x* component of NEO velocity due to impact
P_{ej} = net momentum of ejected material.

Note that the impact and ejection processes are assumed to follow the simple pattern shown in figure 68.

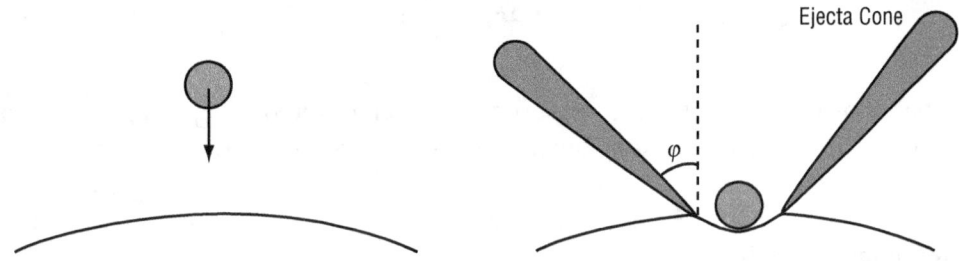

Figure 68. Impact and ejection geometry.

The ejecta material is assumed to emerge from the impact site in the form of a uniform cone; in other words, it all emerges at the same angle (φ) to, and evenly distributed around, the vertical axis.

Assuming that

$$M_{NEO} \gg M_i, M_{ej} , \qquad (60)$$

one can rewrite equation (59) to give

$$\Delta u_x = \frac{M_i v_i}{M_{\text{NEO}}}\left(1 + \frac{P_{ej}}{M_i v_i}\right)\cos\theta \ . \qquad (61)$$

For motion along the y axis,

$$M_i v_i \sin\theta = (M_{\text{NEO}} + M_i - M_{ej})\Delta u_y - P_{ej}\sin\theta \ , \qquad (62)$$

and hence,

$$\Delta u_y = \frac{M_i v_i}{M_{\text{NEO}}}\left(1 + \frac{P_{ej}}{M_i v_i}\right)\sin\theta \ . \qquad (63)$$

Equations (61) and (63) constitute the momentum conservation part of this model. Note that if one neglects the ejecta momentum terms, the model becomes completely inelastic.

If one denotes the cumulative mass of ejecta traveling faster than v by $M(>v)$, then[35]

$$M(>v) = 0.05 R_c^3 \rho \frac{Y}{\rho v^2} \ . \qquad (64)$$

This equation is based on data obtained from experiments that measured the ejecta produced by impact into materials of differing hardness. The various quantities are defined as follows:

R_c = crater radius
ρ = NEO material mass density
v = ejecta speed
Y = NEO material strength.

Defining $v_{\min}$ as the minimum speed of ejected material, the total mass of ejecta can clearly be written as

$$M_{ej} = 0.05 R_c^3 \rho \frac{Y}{\rho v_{\min}^2} \ . \qquad (65)$$

This expression can be used to eliminate Y and R_c from equation (64) to give

$$M(>v) = \left(\frac{v_{min}}{v}\right)^2 M_{ej} \ . \tag{66}$$

Now, consider the small element of ejecta mass which emerges with speed in the range $v \to v + \delta v$. This mass is given by

$$M(>v) - M(>v+\delta v) = \left(\frac{v_{min}}{v}\right)^2 M_{ej} - \left(\frac{v_{min}}{v+\delta v}\right)^2 M_{ej} \ , \tag{67}$$

which can be expanded, retaining only first-order terms in δv, to give

$$\delta M(>v) = 2\frac{v_{min}^2}{v^3} M_{ej} \delta v \ . \tag{68}$$

Noting our assumption that the ejecta is assumed to all emerge at the same angle (see fig. 68), the following expression can be written for the vertical component of momentum (δp) of this small portion of ejecta:

$$\delta p = \frac{2v_{min}^2 M_{ej} \delta v}{v^3} v \cos\varphi \ . \tag{69}$$

This enables one to derive the vertical component of momentum for the entire ejecta mass (P_{ej}) as follows:

$$P_{ej} = \int_{v_{min}}^{\infty} \frac{2v_{min}^2 M_{ej} \cos\varphi}{v^2} dv \ , \tag{70}$$

and hence,

$$P_{ej} = 2v_{min} M_{ej} \cos\varphi \ . \tag{71}$$

Next, an equation that relates the mass of ejecta to the radius of the crater is used.[35] It is derived on the assumption—based on empirical evidence—that the crater is one half of an oblate spheroid with a depth of 0.4 R_c:

$$M_{ej} = \frac{4\pi}{3} \rho R_c^2 \frac{0.4 R_c}{2} \ , \tag{72}$$

which can be rewritten as

$$M_{ej} = 0.8378 \rho R_c^3 \ , \tag{73}$$

from which

$$R_c^3 = \frac{1.1937 M_{ej}}{\rho} . \tag{74}$$

M_{ej} can be eliminated from this expression using equation (65) to give

$$v_{min} = 0.2443 \left(\frac{Y}{\rho}\right)^{0.5} , \tag{75}$$

and hence equation (71) can be rewritten as

$$P_{ej} = 0.4886 \left(\frac{Y}{\rho}\right)^{0.5} M_{ej} \cos\varphi . \tag{76}$$

One final equation is also available.[35] This was originally derived using data from a large number of laboratory cratering experiments and links the total mass of ejecta to the relative speed of collision (V) and interceptor mass as follows:

$$M_{ej} = 0.458 M_i \frac{\rho}{\rho_i} \left(\frac{\rho V^2}{Y}\right)^{0.709} ; \tag{77}$$

thus, M_{ej} can be eliminated from equation (76) and rewritten as

$$\frac{P_{ej}}{M_i v_i} = 0.2238 \frac{\rho}{\rho_i} \frac{V}{v_i} \left(\frac{\rho V^2}{Y}\right)^{0.209} \cos\varphi . \tag{78}$$

This is the final quantity required to solve equations (61) and (63).

Note that the relative speed of collision is given by

$$V = v_i + u\cos\theta . \tag{79}$$

Equation (78) can be used to eliminate ($P_{ej}/M_i \times v_i$) from equations (61) and (63), which can then be used to determine the deflection that results from interceptor impact.

For this method of deflection, the question of fragmentation also needs to be considered. For a relatively low-energy collision, although some material will be ejected from the impact site, the planetary body will remain intact. At larger collisional energies, however, there is the possibility that the body may split into a number of fragments as a result of the impact. It is important to be able to understand and model the fragmentation process, as it may determine the usefulness of kinetic deflection as a threat mitigation technique. An incoming body that is unexpectedly fragmented by a kinetic deflection interceptor may subsequently prove very difficult to deal with if some of the fragments are very massive and still pose a threat.

There is clearly very limited direct experimental data to draw on. Studies have been conducted to determine the effect of kinetic energy on various NEO analogue materials.[36] The mechanics of collision and fragmentation are reasonably well understood for impact speeds of up to ≈8 km/s. Understanding is less clear for impact speeds in the 8 to 15 km/s range. At speeds >15 km/s, there is very poor understanding, which is unfortunate, given that this is the likely regime in which a kinetic deflection interceptor would operate.

In general terms, it is understood that the initial impact produces a "hydrodynamically induced" crater; in other words, the cratering process takes place while the solid material of the body has been rendered fluid-like by the impact energy. The impact produces, on a timescale of between microseconds and milliseconds, internal shock waves that propagate through the target on a timescale of between tens and thousands of milliseconds. Depending upon the impact energy and the constitution of the target body, the final outcome will range from velocity deflection to complete and catastrophic fragmentation.

The critical parameter in determining whether the target will fragment is the collisional energy per unit mass of the target body. This is written as E_P/M_T, where E_P is the collisional energy and M_T is the target mass. The following approximate criteria has been established for catastrophic fragmentation;[36] i.e., where the target is rendered into a large number of fragments, each much smaller than the original body:

- For $E_P/M_T < 0.5$ J/gm, the target is considered to be cratered but intact.
- For $E_P/M_T > 0.5$ J/gm, the target is considered to be catastrophically fragmented.

Some empirical data[36] are available that relate the collisional energy to the mass of the largest fragment produced by the collision. The data are reproduced in table 23 for two different target materials, showing the relative size of the largest fragment (M_L/M_T) at various values of (E_P/M_T). Note that M_L denotes the mass of the largest fragment.

Table 23. Relative size of largest fragment at various collisional energies.

Target Material	E_P/M_T (J/gm)	M_L/M_T	E_P/M_T (J/gm)	M_L/M_T	E_P/M_T (J/gm)	M_L/M_T
Rock (basalt)	0.07	0.9	3	0.1	10	0.01
Ice	0.01	0.9	0.05	0.1	0.6	0.01

For facilitating computations, these data can be roughly curve fitted as shown in figure 69.

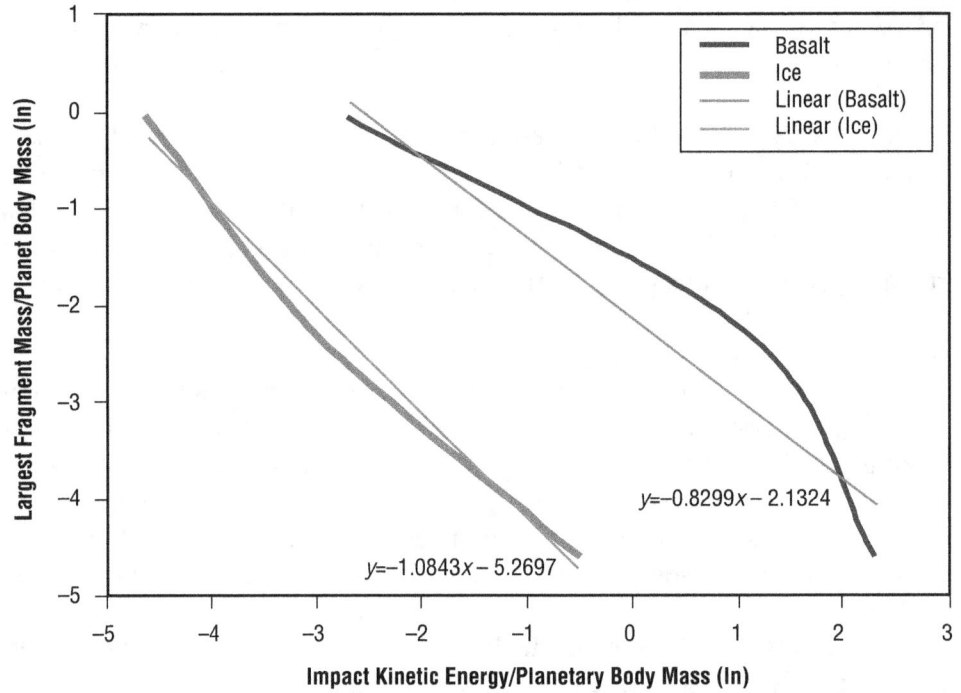

Figure 69. Fragmentation data and curve fit results.

The curve fit results can be summarized as follows:

$$\ln\left(\frac{M_L}{M_T}\right) = A \ln\left(\frac{E_P}{M_T}\right) + B \ , \tag{80}$$

where

$A = -0.8299$ for basalt; -1.0843 for ice.
$B = -2.1324$ for basalt; -5.2697 for ice.

In conclusion, some indications[36] are noted that current analytical and semiempirical models may underestimate the momentum change resulting from a high-velocity impact. Chemical reactions, taking place in the fluid phase after impact, are probably responsible for this discrepancy. Hence, the above data and model should be viewed as very preliminary. This area requires more work, including, if possible, actual impact tests on a variety of planetary bodies.

Table 24 lists the qualitative considerations that affect the use of kinetic deflection. Small quantities of dust surrounding the target body would not pose a major problem provided the interceptor is designed appropriately. However, larger pieces of debris could pose a serious problem, destroying it before it impacts on the target. Unlike some of the other techniques, kinetic deflection works well with a rotating body. In fact, a rotating target is in some respects more desirable than a nonrotating target as it will be easier to identify the location of its center of mass.

Table 24. Qualitative considerations for threat mitigation using kinetic deflection.

First-Order Qualitative Considerations	
Sensitivity to dust cloud	Low
Ability to handle target rotation	High
Requires landing on target	No
Usefulness on fragmented body	Medium
Swarm option	High
Second-Order Qualitative Considerations	
Usefulness as weapon	High
Perceived safety	High
Synergy with other NASA missions	
Manned missions	Low
Robotic missions	Low
Resource utilization missions	High
Costs	
Development	Low
Deployment	Low

Perceived safety is judged to be high because of the lack of any sort of nuclear element to this technique. Technology developed for ballistic missile interception will probably be directly applicable here, which should reduce development costs. Note that the closing speed for this application may be an order of magnitude higher than for missile interception. Although potentially usable as a weapon itself, a kinetic deflection system would be more likely to be employed in a defensive capacity than in an offensive one.

In the present application, it is not anticipated that a single interceptor would be employed, but rather a stream of them, each capable of some terminal-phase maneuvering. The later interceptors in a stream will have been launched long before the initial interceptors impact, so they must be able to maneuver in order to respond to changes in the body's trajectory.

6. TRAJECTORY MODELING

6.1 Outbound

The outbound trajectory was modeled simply using two-body orbital mechanics and impulsive thrust assumptions. These assumptions are not accurate for the continuous thrust propulsion systems considered, but the use of a more accurate integrating trajectory optimization program would have required more time than was available for this project. These inaccuracies must be considered in any follow-on study.

The outbound trajectory is solved as a Gauss problem. Two points in space are known, as well as the desired transfer time between them. The asteroid's position at interception or rendezvous is calculated by assuming an Earth impact position and traveling backward along the asteroid's orbit to the desired arrival time, which is given as the number of days before impact. In this study, the asteroid's orbit is initialized such that it will impact the Earth at a 45° angle on the heliocentric-ecliptic plane. The spacecraft's position at departure is the same as the Earth's at that time. The Earth's position is calculated by moving the Earth backward from the impact point by the number of days equal to the asteroid arrival time—given as the number of days before impact—plus the desired outbound trajectory flight time.

The Gauss problem formulation used in this study is taken from the literature.[37] The universal variables solution method allows the trajectory to be any type of conic section: an ellipse, a parabola, or a hyperbola. Two ΔVs are calculated: (1) Must be applied to depart Earth's orbit and send the vehicle on the trajectory that will intercept the asteroid in the desired flight time, and (2) applied upon arrival at the asteroid. This ΔV places the vehicle in the asteroid's orbit and allows for rendezvous with the asteroid. Interceptor missions like kinetic deflection use only the first ΔV, since impact with the asteroid is desired. Other missions that require close asteroid operations must perform both maneuvers.

The calculations were implemented in a Mircosoft® Excel workbook and then "wrapped" into the integrated design environment, ModelCenter®. All outbound trajectories are designed to intercept or rendezvous with a baseline NEO that is defined in section 6.2. Example results for missions to the baseline asteroid are given in figure 70. The ΔVs for both rendezvous and interception are given. These results correspond to a total mission time of 3,600 days. Total mission time is defined as the sum of the outbound flight time and the asteroid arrival time (days before impact), or the time between launch of the system and the asteroid's predicted collision with the Earth.

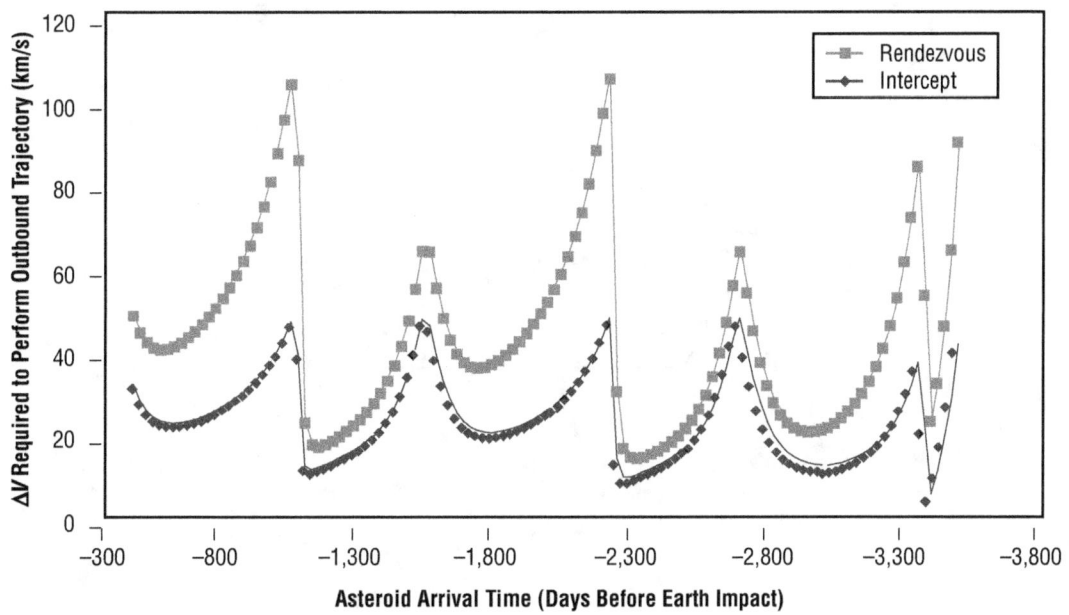

Figure 70. Outbound trajectory ΔVs for 3,600-day total mission duration.

6.2 Inbound

The inbound trajectory modeling software determines the minimum impulsive ΔV required to make an incoming planetary body miss Earth by some specified distance. In order to allow the inclusion of other objects, such as Jupiter, during later studies, the software numerically integrates the equations of motion of the Sun-Earth-planetary body system. This program, Planetary Body Intercept (PBI), iterates over the search space until a ΔV is found that is a minimum and also causes the planetary body to miss the surface of Earth by 3 Earth radii.

To determine minimum ΔV requirements, it was necessary to find a planetary body that would definitely collide with Earth, preferably dead center. However, due to the uncertainty in the orbital determination of the NEOs, it was decided not to conduct a lengthy search of the NEO catalog. Instead, a fictitious body was created in the following way. First, the orbital elements of 444 known potentially hazardous asteroids (PHAs) were examined in order to establish those elements that might apply to an "average" PHA. Next, the PHA database was searched for one asteroid that came close to this average—at least in terms of orbital size, eccentricity, and inclination. The resulting candidate asteroid was 1999JT6. Its orbital elements were then modified slightly, so as to force a collision with Earth. For this purpose, the Earth was placed 45° from the x axis of the Heliocentric-Ecliptic coordinate system at the time of the hypothetical impact, as illustrated in figure 71. The original and modified elements of 1999JT6 are given in table 25. The modified asteroid orbit is plotted in figure 72.

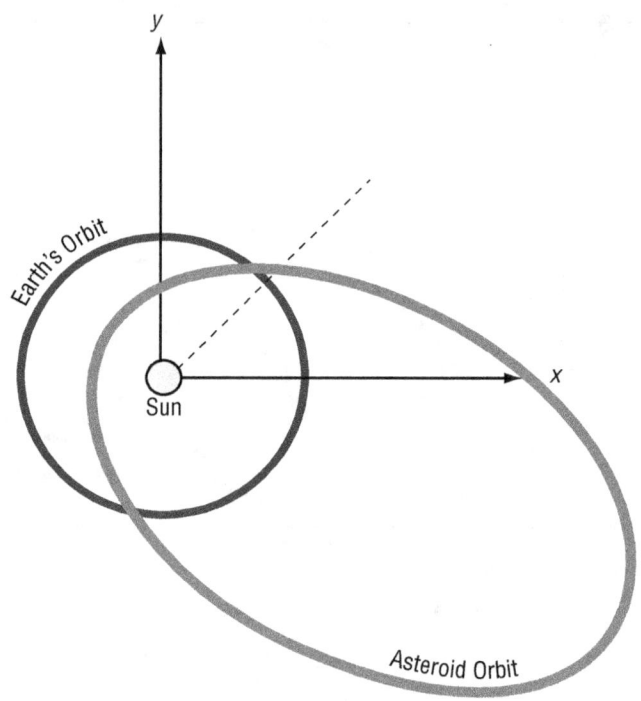

Figure 71. Illustration showing a typical NEO orbit. The velocity of the planetary body at impact for this case is $(-40,0,0)^T$ km/s, parallel to the x axis of the Heliocentric-Ecliptic system.

Table 25. Original and modified orbital elements of 1999JT6.

Orbital Elements	Original	Modified
Name	1999JT6	M1999JT6
Semimajor axis (km)	319280491.2	319285502.2
Eccentricity	0.579033744	0.5791277
Inclination*	9.568048	11.47182
Longitude of ascending node*	79.06928886	45.009263
Argument of perihelion*	38.8692997	41.840244

*All angles are in degrees.

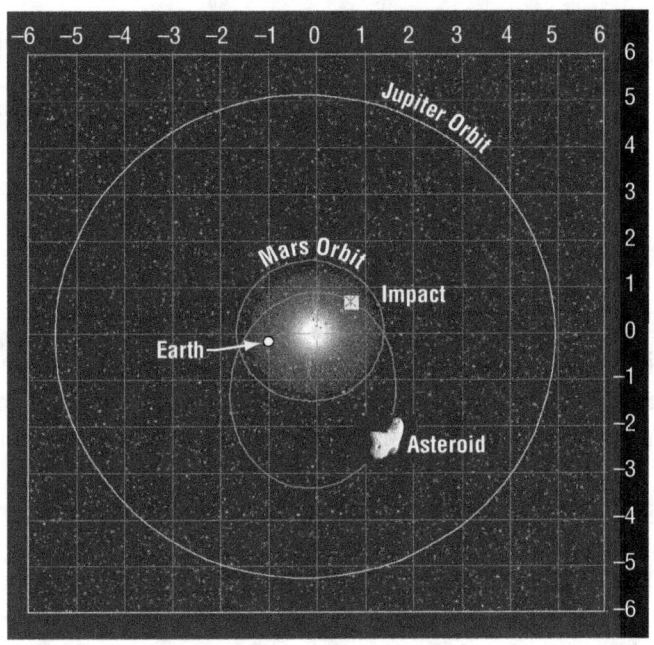

Figure 72. M1999JT6 orbit plot.

To avoid confusion with the real asteroid 1999JT6, this fictitious asteroid has been named M1999JT6. This step was taken largely to avoid the possibility of readers getting the impression that 1999JT6 is indeed on a collision course with the Earth. Size and composition of the fictitious asteroid are not specified, since these characteristics are varied during the analysis. During execution of the PBI program, the user specifies only the mass and velocity vector of an arbitrary planetary body, which is initially placed at the center of the Earth. Then, the position of the body is adjusted so that it collides with the surface of the Earth nearly dead center. For M1999JT6, the mass is varied, but the impact velocity vector is always as given:

$$V = -35.65I + 18.5J + 9.0K \text{ km/s} . \tag{81}$$

This impact velocity matches the modified orbital elements above, which are determined when the asteroid is at a distance from Earth equal to twice the radius of the Earth's sphere of influence.

The user also specifies the number of days before impact when the deflection ΔV is to be applied. Backward numerical integration of the three-body equations (eq. (82)) positions both the Earth and planetary body at their proper location for this specified time. The equations are written in the Heliocentric-Ecliptic coordinate system, with the bodies assumed to be point masses:

$$\ddot{\mathbf{r}}_{E/S} = -\frac{G(M_E + M_S)}{\|\mathbf{r}_{E/S}\|^3}\mathbf{r}_{E/S} + GM_{PB}\left[\frac{\mathbf{r}_{PB/E}}{\|\mathbf{r}_{PB/E}\|^3} - \frac{\mathbf{r}_{PB/S}}{\|\mathbf{r}_{PB/S}\|^3}\right]$$

$$\ddot{\mathbf{r}}_{PB/S} = -\frac{G(M_{PB} + M_S)}{\|\mathbf{r}_{PB/S}\|^3}\mathbf{r}_{B/S} - GM_E\left[\frac{\mathbf{r}_{PB/E}}{\|\mathbf{r}_{PB/E}\|^3} + \frac{\mathbf{r}_{E/S}}{\|\mathbf{r}_{E/S}\|^3}\right]. \quad (82)$$

At this point, the iterative process of determining the magnitude of the impulsive ΔV begins. The direction of the impulse is determined by the user from a list of possible commands shown in table 26. Each maneuver is made from the perspective of the planetary body and its orbit, not from the perspective of Earth's orbit. The program makes an initial guess for the magnitude of the ΔV and integrates forward until the planetary body reaches its point of closest approach to the Earth. If the minimum distance between Earth and the planetary body is not within some user-specified tolerance, a new ΔV magnitude is chosen and the process repeated. In practice, with a tolerance of 0.01 Earth radii, the program usually converges to a solution after three to five iterations. Some cases, however, require many more iterations, or may not converge at all, as is the case when the planetary body is so close to Earth that no ACCEL maneuver can result in a miss. In these cases, the program prints a warning message and reports the magnitude of the ΔV to be 1×10^{10} km/s.

Table 26. Explanation of the different maneuvers available for use in the program PBI.

Maneuver	Description
ACCEL*	Increase the magnitude of the velocity; do not change direction
DECEL*	Decrease the magnitude of the velocity; do not change direction
UP	Rotate the velocity vector up; do not change the magnitude of the velocity
DOWN	Rotate the velocity vector down; do not change the magnitude of the velocity
INSIDE	Rotate the velocity vector toward the inside of the body's orbit, or toward the Sun, whether or not the orbit is direct or retrograde; do not change the magnitude of the velocity
OUTSIDE	Rotate the velocity vector toward the outside of the body's orbit, or toward the Sun, whether or not the orbit is direct or retrograde; do not change the magnitude of the velocity

*Only ACCEL and DECEL change the magnitude of the velocity vector.

Different maneuvers usually result in different ΔV requirements, and a maneuver that is best in one situation is not always the best in others. For example, for long-period comets coming in at very high speeds, UP or DOWN may result in the minimum ΔV requirement when impact is only a few days away, as shown in figure 73. Note that the velocities shown in figures 73–77 are those of the planetary body at impact. However, figure 74 shows that DECEL may result in the minimum ΔV when the impact is from 50 to 150 days away. The situation is different for the typical asteroid, which would be moving much slower than a long-period comet. Figure 75 shows that the UP maneuver no longer is the best option when the planetary body with an impact velocity of 35 km/s is only a few days away; now the best option is to decelerate the object. With more time available, the preferred option changes from DECEL to OUTSIDE, as illustrated in figure 76. Overall, however, the ACCEL maneuver is found to be the most efficient maneuver for rendezvous times of ≈300 days or more, as illustrated in figure 77.

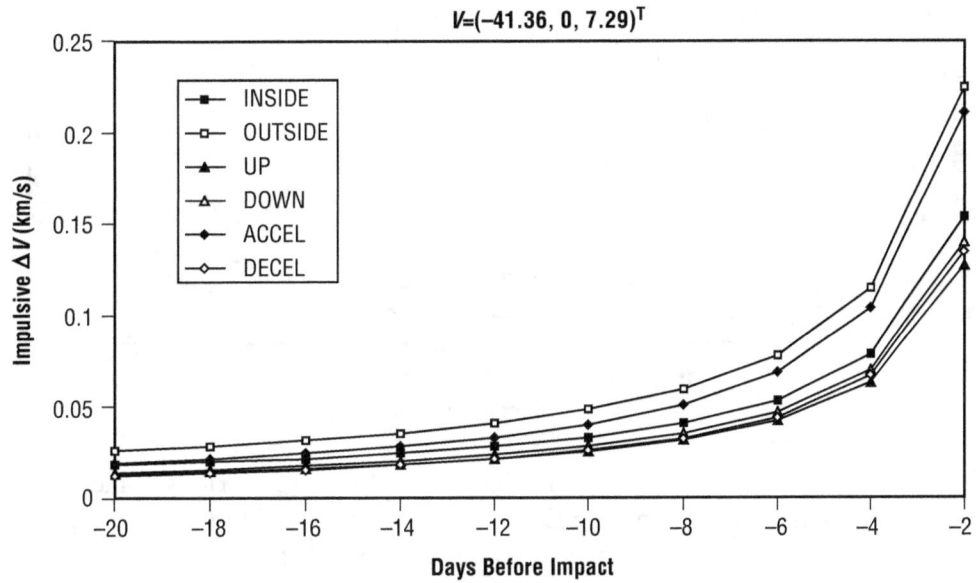

Figure 73. Required impulsive ΔV for 42 km/s velocity for various maneuvers to avoid collision with Earth, showing the benefit of the UP maneuver when impact is only a few days away.

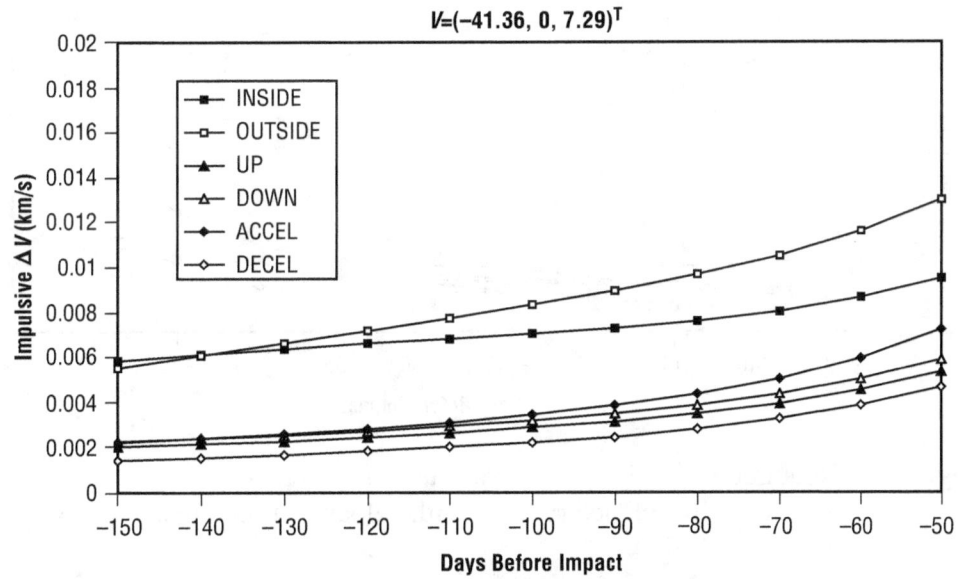

Figure 74. Required impulsive ΔV for 42 km/s velocity for various maneuvers to avoid collision with Earth, showing the benefit of the DECEL and OUTSIDE maneuvers when impact is several weeks away.

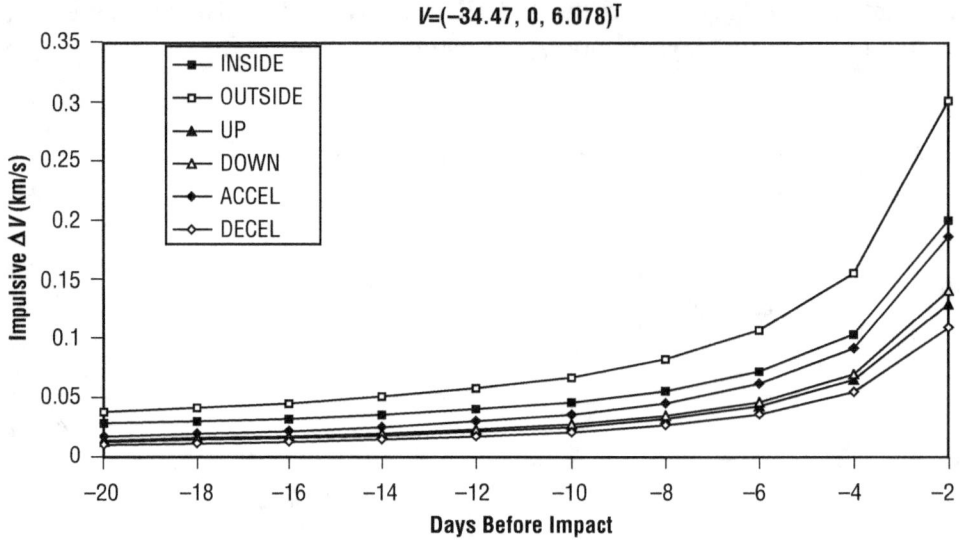

Figure 75. Required impulsive ΔV for 35 km/s velocity for various maneuvers to avoid collision with Earth, showing the benefit of the DECEL maneuver when impact is only a few days away.

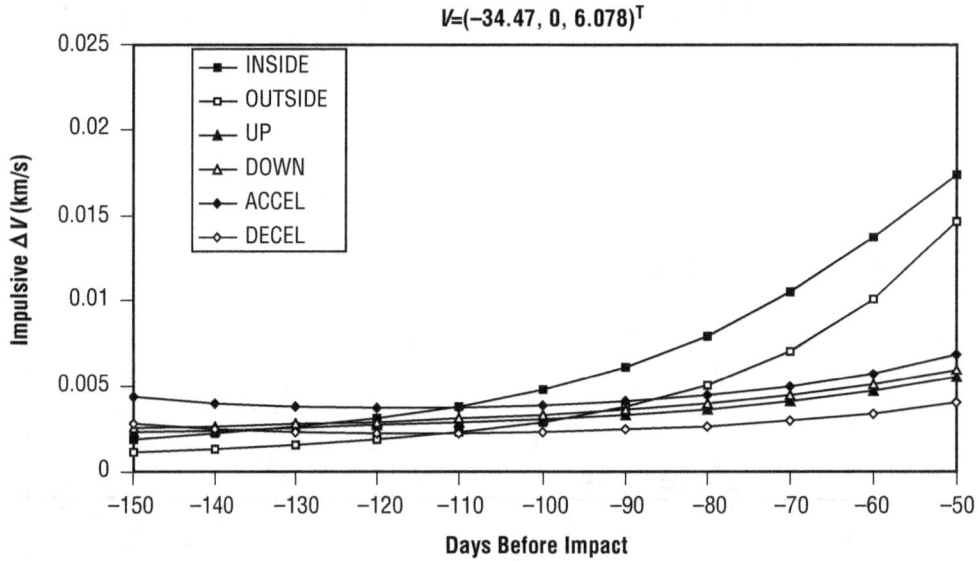

Figure 76. Required impulsive ΔV for various maneuvers to avoid collision with Earth for planetary body with velocity of 35 km/s.

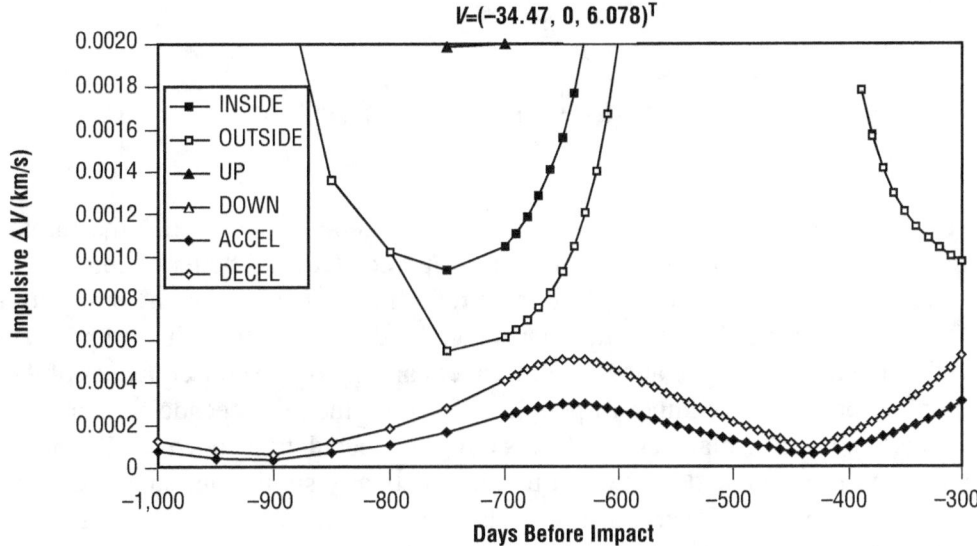

Figure 77. Required impulsive ΔV for various maneuvers to avoid collision with Earth for planetary body with velocity of 35 km/s (long lead time).

By limiting the number of possible maneuvers to six rather than performing a global search, computation speed is increased significantly. In many cases, for example, the optimum trajectory may be a combination of UP and DECEL. However, comparing the ΔV requirements for a range of successful trajectories gives insight into which combinations could possibly result in a lower impulse requirement. The goal of manipulating the planetary body's trajectory is to miss the moving Earth, not to rendezvous with it. Therefore, maneuvers that delay the arrival of the planetary body generally allow Earth to move out of the way before the body arrives.

One must realize that the quest for the minimum ΔV did not include the object's disposal; i.e., never threatening Earth again, only its avoidance. By only requiring the body to miss Earth on this one occasion does not guarantee that it will miss Earth during some future encounter. A maneuver that removes the body completely from the list of threatening objects is clearly the best option and should be considered in future studies.

7. THREAT PARAMETRIC

Thanks to the pioneering work of researchers like Eugene Shoemaker,[38] the threat and consequences of an impact with a planetary body are now more appreciated, and better understood, than was the case 30 yr ago. One only has to consider the literature in references 39–44 and many others to see evidence of increased efforts to understand the threat that Earth faces every day. Although our understanding of the impact threat is still incomplete, it is far ahead of our understanding of the consequences of the most likely impacts. Attempting to predict the number of people killed over the next decade, century, or millennium due to impacts of certain sizes of planetary bodies is a highly speculative endeavor. Recorded impacts in the developed world are rare and so do not constitute a statistically significant database. This means that researchers must resort to the use of theoretical models[41] to estimate the number of deaths resulting from the impact of a planetary body of a certain size, velocity, and type.

Our program, PEOPLE, estimates the number of people saved over the next century if all planetary bodies of a given type; i.e., chondrite, long-period comet, etc., having kinetic energy less than or equal to some given value, can be successfully deflected. The number of fatalities prevented is based on the work done by Shoemaker,[38] Chapman and Morrison,[39] and Lewis[41] using estimates of both the impact frequency and the number of deaths due to impacts of a certain energy. Over 10,000 runs of a modified version of John Lewis's Monte Carlo simulation program were used to generate data for the average number of deaths due to each type of object (table 27). (The main modification to Lewis's simulation code allowed the tallying of deaths due to different types of objects, enabling the study team to focus efforts on the types of objects that would most likely cause the most deaths over the next century.)

Table 27. Types of planetary bodies examined in the Monte Carlo simulation and their average contribution to the total number of deaths over the next century.

Type	Total Deaths (%)
Chondrite	60
Achondrite	5
Iron	5
Mesosiderite	<1
Pallasite	<1
Comet, short-period	6
Comet, long-period	24

The program PEOPLE also determines which parameter—mass or velocity—should be increased to counter the largest portion of the threat. This determination is based on velocity distributions from Lewis[41] and Chesley,[43] and size distributions from Gold[40] and Ivezic et al.[44] (Recent data from Brown et al.[45] indicates that the frequency for Tunguska-sized events may occur only every 1,000 yr as opposed to previous estimates of every 200 to 300 yr, and that the size distribution of the smaller asteroids may need to be reassessed.) A static human population of 6 billion is assumed here, as it is throughout most of the literature. Causes of death include tsunamis, blast waves, firestorms, and direct impacts.

The first step in PEOPLE is to use the mass (m) of the planetary body (kilograms), its impact velocity (V) relative to Earth (m/s), and its type (see table 27), and determine the equivalent energy yield in megatons of TNT using the following equation:

$$\text{Mton} = \frac{mV_{PB/E}^2}{8.37 \times 10^{15}} . \tag{83}$$

Next, the number of deaths (F) per year, on average, due to all impacts having this energy or less is determined. These equations were derived using data taken from Chapman and Morrison,[39] Lewis,[41] and Gold.[40] The equations are split into four categories based on the planetary body's equivalent energy yield:

$$\begin{aligned}
&\text{Mton} \leq 2.5 & &F = 0.0236 \text{ Mton}^{5.525} \\
&2.5 < \text{Mton} \leq 200{,}000 & &F = 164 - 180 \text{ Mton}^{-0.1333} \\
&200{,}000 < \text{Mton} \leq 1 \times 10^8 & &F = 3763 - 4.16 \times 10^6 \text{ Mton}^{-0.577} \\
&\text{Mton} > 1 \times 10^8 & &F = 3763 - 4.16 \times 10^6 \text{ Mton}^{-0.577} \\
& & &\text{(if the user wishes; this is an extrapolation).}
\end{aligned} \tag{84}$$

PEOPLE also calculates the number of fatalities (f) that would have been expected on average for this energy of impact. This is not the cumulative number of fatalities as described above, but is based on the following equations:

$$\begin{aligned}
&\text{Mton} \leq 2.5 & &f = 5.433 \text{ Mton}^{6.325} \\
&2.5 < \text{Mton} \leq 200{,}000 & &f = 1000 \text{ Mton}^{2/3} \\
&200{,}000 < \text{Mton} \leq 1 \times 10^8 & &f = 1 \times 10^8 \text{ Mton}^{0.223} \\
&\text{Mton} > 1 \times 10^8 & &f = 6 \times 10^9, \text{ the world population.}
\end{aligned} \tag{85}$$

Caution must be used when applying these equations and estimates. For example, the overwhelmingly dominant event each year is a small one, which would result in only a few fatalities. However, while a catastrophic event with impact energy of 1×10^8 Mton could kill several billion people, its likelihood is only 1 in 100 million.[39,43] Therefore, this type of impact will result in <100 fatalities per year on average, but when it does occur, the outcome will clearly be catastrophic. Similarly, while their likelihood is only 1 in 10,000, impactors with energies of 1×10^5 Mton statistically cause ≈3,000 fatalities per year, the largest number by far. Impacts of this size are on the threshold of being globally catastrophic events, large enough to do massive damage throughout the world, but too small to destroy humanity. They are also frequent enough to result in a high average number of fatalities per year.[39]

Finally, if the deflection system used to successfully change the course of the object has some excess energy, the program PEOPLE determines which planetary body parameter—velocity or mass—should be increased in order to defeat the maximum portion of the total threat. These equations are based on data reported by Lewis[41] and Gold[40] on the velocity and size distributions of NEAs and comets. First, the spherical diameter of the object is determined by

$$D = \left(\frac{6m}{\pi \rho}\right)^{1/3} , \qquad (86)$$

where the density depends of the type of object, as listed in table 27. The rate of change of cumulative average fatalities with respect to mass is then calculated from

$$\frac{\partial F}{\partial m} = 37.67 D^{-2.42241} . \qquad (87)$$

Finally, the program estimates the rate of change in fatalities with respect to velocity, using the appropriate form of equation (88). This calculation depends on the type of object (long-period comet, short-period comet, or asteroid) and its velocity relative to the Earth at impact:

- Long-period comet:

$$11.18 \leq V < 21.18 \qquad \frac{\partial F}{\partial V} = 0.000075(V - 11.18)^2$$

$$21.18 \leq V < 35 \qquad \frac{\partial F}{\partial V} = 0.0075$$

$$35 \leq V < 50 \qquad \frac{\partial F}{\partial V} = 0.0015V - 0.045$$

$$50 \leq V < 66.86 \qquad \frac{\partial F}{\partial V} = 0.03$$

$$66.86 \leq V < 71.86 \qquad \frac{\partial F}{\partial V} = -0.005256V + 0.3814$$

- Short-period comet:

$$14 \leq V < 34 \qquad \frac{\partial F}{\partial V} = -0.00072(V-24)^2 + 0.072$$

$$34 \leq V < 66.86 \qquad \frac{\partial F}{\partial V} = 0$$

$$66.86 \leq V < 71.86 \qquad \frac{\partial F}{\partial V} = 0.008$$

- Asteroid:

$$11.18 \leq V < 71.86 \qquad \frac{\partial F}{\partial V} = 0.1023(V-11.18)^{-0.79} \: . \tag{88}$$

The velocity distribution of asteroids is based on an equation slightly different from the one used by Lewis,[41] but it seems to fit the data slightly better.

PEOPLE was incorporated into a model in ModelCenter to determine the number of deaths that could, on average, be prevented over the next century if all incoming asteroids of a particular energy or less could be successfully deflected. The results are graphed in figure 78. The threshold asteroid size for a globally catastrophic event is evident, although the exact size of the asteroid that defines the threshold is unclear, and is strongly influenced by type.

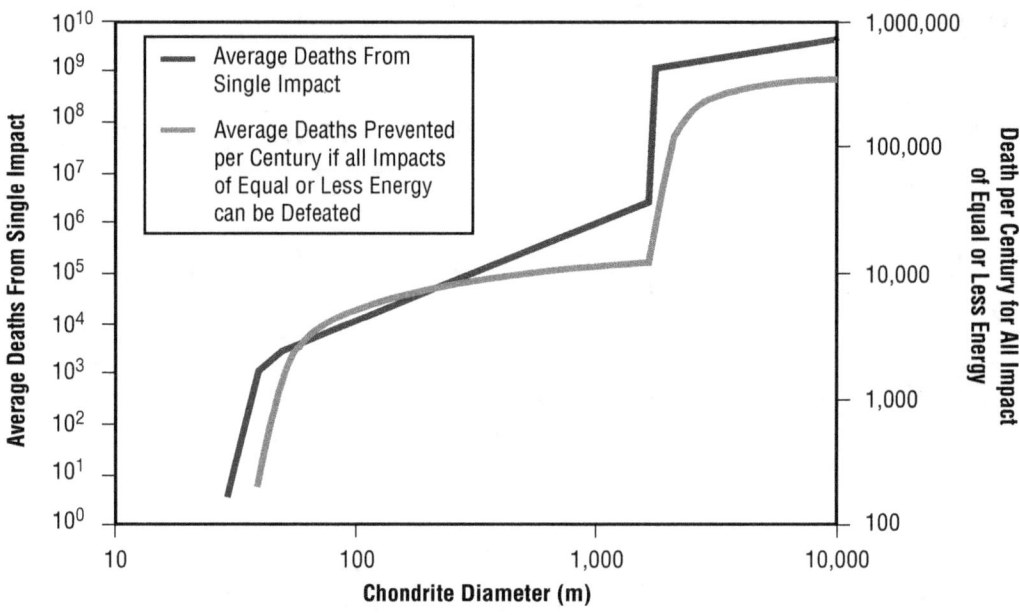

Figure 78. Average deaths from single asteroid impact versus size, and the average total number of deaths prevented if all impacts of equal or less energy can be avoided.

The number of fatalities is very sensitive to the impact location for the smaller asteroids. The average number of fatalities from the impact of a 100-m chondrite is ≈10,000 people, but that number could increase to the millions if the impact occurred in or near a major city or a densely populated coastline. Therefore, the relatively small average number of fatalities per century is little reason to conclude that defending Earth from these sizes of impacts is unnecessary.

In summary, the average number of deaths per impact and the cumulative number of deaths per century are based on simulations. These simulations contain both statistical data and estimates gleaned from experience with high-energy detonations in both populated and unpopulated areas. This results in a high uncertainty in the estimate of fatalities due to an asteroid impact. An uncertainty analysis of the estimated fatalities was not completed due to time constraints. However, it is clear that the standard deviation in the estimate of facilities for most sizes of impactors is much larger than the mean. Therefore, the estimate of average fatalities per century should be used with caution, and is only a guide to illuminate the direction in which the design of a planetary defense system should proceed in order to prevent the most fatalities on average.

8. PARAMETRIC RESULTS

8.1 Integrated Analysis

The original intent of this project was to evaluate the ability of various combinations of technologies to defeat the entire threat posed by NEOs. Obviously, this is a very complicated problem. Potential impactors come in all shapes and sizes, and their orbits vary greatly. To further complicate matters, a large number of technologies for use in threat mitigation have been examined. Assessing all possible technology combinations would be prohibitive.

The original intent of this project was to select several technologies based on our understanding and experience, and to test the ability of each to defeat the total threat. Figure 79 illustrates the proposed analysis process. Starting with an assumption for the total system mass and the total mission time allowed, the analysis process then divides, based on the type of threat mitigation concept being considered—remote station, interception, or deflection. The remote station analysis path assumes both an incoming asteroid mass and a velocity vector. Running the inbound parametric defines the ΔV required to deflect the asteroid, given its size and velocity vector at point of impact, and also the allowable mission time. Running the remote station tool, based on the ΔV to be delivered to the object and the allowable mission time, computes the required remote station mass. If this mass is not equal to the allowable mass for the system assumed at the beginning, then the analysis path returns to assume a new asteroid mass and velocity vector. New ΔVs and remote station masses are then computed. The new asteroid mass and velocity vector is selected using the threat assessment tool to maximize the percentage of the total threat that can be defeated for the assumed total system mass. After closure, the threat assessment tool is run again in order to compute the total threat that is defeated. The total threat is quantified by the percentage of people saved by deployment of the system over a given time period divided by the number of people expected to die from impact of an NEO over the same period. By running through this process several times—assuming new total system masses and mission times on each occasion—yields a parametric model of the total threat defeated as a function of total system mass for lines of constant mission time.

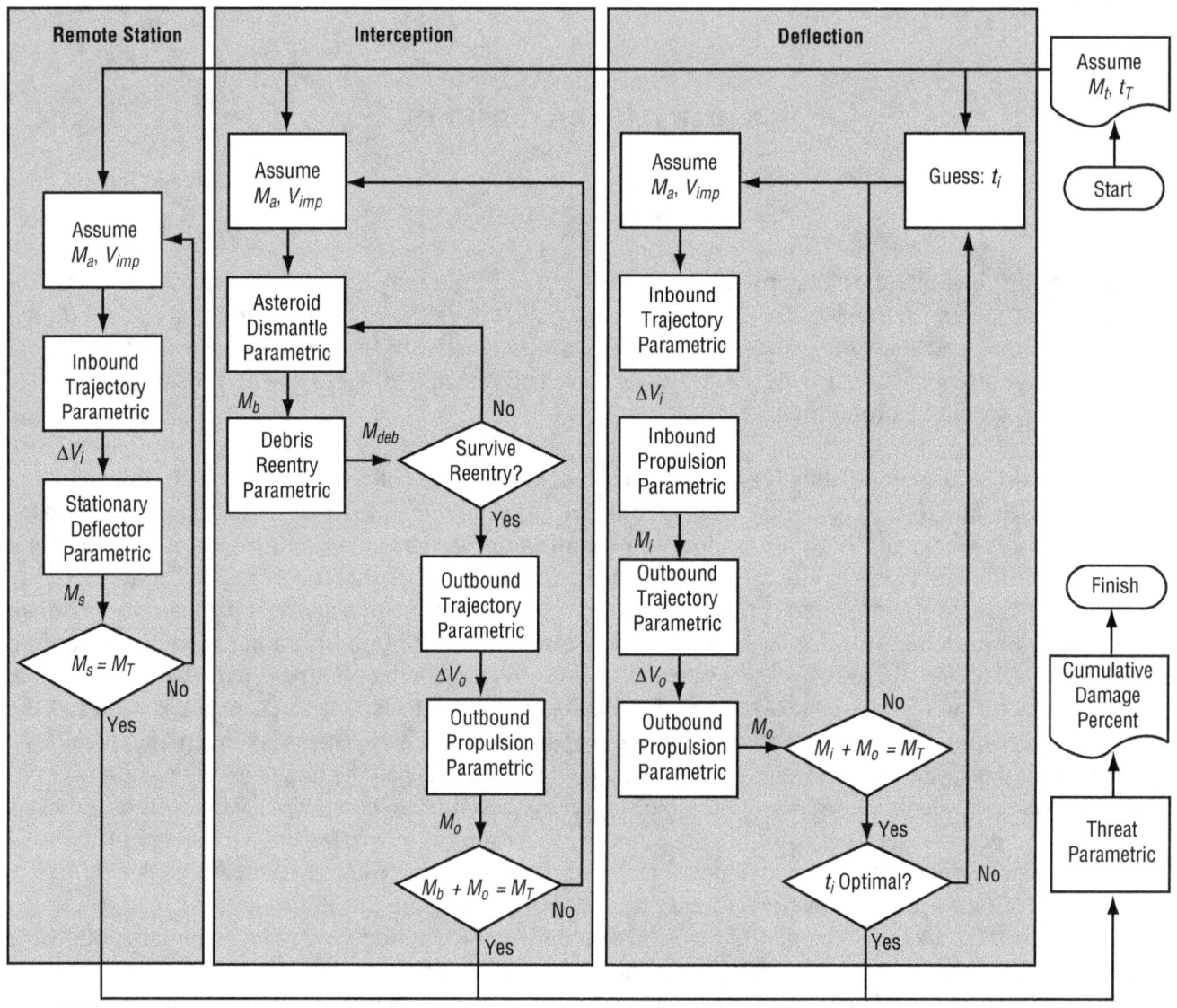

Figure 79. Proposed analysis process for assessing total amount of threat mitigated.

The interception branch differs from that of the remote station in that the inner iterative loop determines the asteroid size and velocity that can be deflected. Here, there is an additional problem in optimizing the amount of the total mission mass allocated to the outbound propulsion against the amount allocated to the interception system. First, an interceptor mass is assumed and the size of the resulting object fragments is estimated. Development of an atmospheric entry code that would model the burnup of these fragments was initiated, but this has not yet been completed. It is intended that this model would include the effects of drafting; i.e., later objects following in the wake of the earlier objects. If the fragments survived reentry, then a larger interception mass is assumed and a new iteration ensues. Otherwise, the calculation proceeds to calculate the ΔV required and then the mass of the outbound propulsion stage needed to take the interceptor to the incoming object. This outer loop iterates until the maximum threatening object that can be mitigated for the total allowable mission mass is found, again using the threat tool. As before, a parametric model, giving the percentage of the total threat defeated as a function of both total system mass and mission time, can be generated.

The rendezvous branch in figure 79 is the most complicated one of all. Here, the most threatening object must be found using techniques similar to those employed for the interception branch. However, in this case, an allowed inbound trip time is also assumed. The analysis process then runs through the tools that calculate the required inbound ΔV and mass, as well as the outbound ΔV and mass. Then, the total mass can be calculated by summing the inbound and outbound masses and comparing the result with the assumed total system mass. Even after closure of this inner loop, the interception point, defined by the allowed inbound interception time, may not be optimal; therefore, another loop is used to find the optimal interception point. After closure of this loop, the process goes into the threat parametric to find the total percentage of the threat that is defeated. Again, a parametric model of total threat defeated as a function of system mass and mission time is generated.

There are still two assumptions built into the above analysis scheme:

(1) Although the distribution of object mass and velocity is taken into account, the possible distribution in composition is not. That composition is believed to be a secondary factor in performance, although not in damage caused. In addition, because asteroid and comet compositions are so poorly understood, it was decided that the whole issue of composition would not be addressed for this initial study. Note that the inclination distribution was taken into account; it is imposed by a nonzero z component in the incoming object velocity vector at impact.

(2) The deflection study allowed no time for setup after asteroid rendezvous. None of the options considered required a significant amount of time after rendezvous for these operations, except for the mass driver, so it was assumed that the required deflection ΔV was imposed instantaneously upon rendezvous. Finally, the threat parametric has several implied assumptions defined in section 7.

Unfortunately, these ambitious analysis goals were not completed in the time available. As will be shown below, a parametric model of total system mass was derived for several architecture options, assuming a standard set of orbital elements for the incoming object. The parametrics were derived using a process similar to the inner loops shown in figure 79. The architecture options considered and performance of these options are described in section 8.2.

8.2 Architecture Options

In the time available for this study, several architecture options were considered—either suggested in the literature or which appeared promising. It must be emphasized that this selection does not constitute a full list of possible options for threat mitigation. In particular, note that a remote station architecture option was not considered. The list of cases that were considered and their mission configuration are shown in table 28. Note that each case or mission scenario was assembled in a unique project file within the integrated design environment ModelCenter.

Table 28. Architecture options considered in this study.

Outbound System	Inbound System	Remote Station Versus Interception Versus Rendezvous	Deflection Versus Fragmentation
Staged chemical	Mass driver	Rendezvous	Deflection
Staged chemical	Kinetic deflector	Interception	Deflection
Staged chemical	Nuclear deflection	Interception and rendezvous	Deflection
Nuclear pulse	Nuclear deflection	Rendezvous	Deflection
Solar collector	Solar collector	Rendezvous	Deflection

8.3 Parametric Performance

8.3.1 Staged Chemical/Mass Driver

The basic design process for this scenario is as follows. The inbound trajectory program is initiated by specifying how many days before impact the deflection is to take place. One output of this tool is the ΔV required to deflect the asteroid. This value is used in the mass driver sizing tool. The asteroid's position at the time of deflection is another output. With this information and the desired time of flight, the outbound trajectory tool determines the ΔV requirement to rendezvous with the asteroid. The mass driver's total mass is used as the payload mass for a staged chemical rocket that performs the ΔV maneuvers required for the outbound trajectory. The final result is the total system mass, which includes the mass driver and the rocket required to deliver it to the asteroid. The optimal mission from an energy standpoint is that which requires the lowest mass. This requires a balance between the conflicting goals of minimizing the deflection ΔV and the outbound trajectory ΔV.

Figure 80, a partial screen shot from the ModelCenter program, shows this basic design process. The first component, Main, is a simple script. The primary reason for creating Main is to collect all of the inputs used in the parametric studies into a single component. This frees the user from needing to search through the extensive input lists of the other programs. It also performs some simple algebraic manipulation necessary to convert units and calculate inputs for subsequent programs.

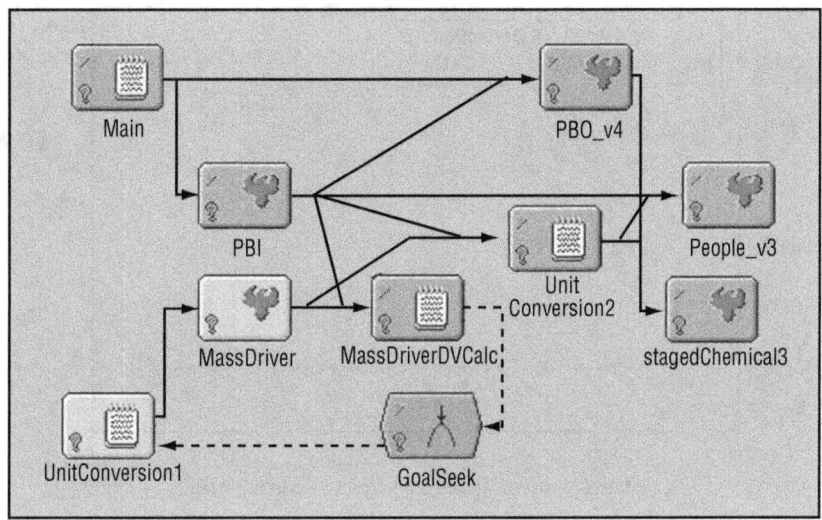

Figure 80. Staged chemical/mass driver model.

The next component, PBI, is the inbound trajectory tool, which was documented in section 6.2 of this TP. With the asteroid specified and the integration parameters set, the only changing input to this program is the number of days before impact that the deflection is to occur, and the optimal deflection direction. The first of these inputs is a user input that comes from the Main program discussed above. The second input is determined by looking at each deflection direction and finding the best option for the missions considered. Figure 81 shows the deflection ΔV for each deflection direction over the range of 100 days to over 10 yr before impact. In the magnified image, it is apparent that for any deflection that occurs more than 600 days before impact, the ACCEL maneuver requires the least ΔV. It is also interesting to note that the ΔV required oscillates with each asteroid period, and it is obvious for the ACCEL and DECEL options that each oscillation diminishes in magnitude.

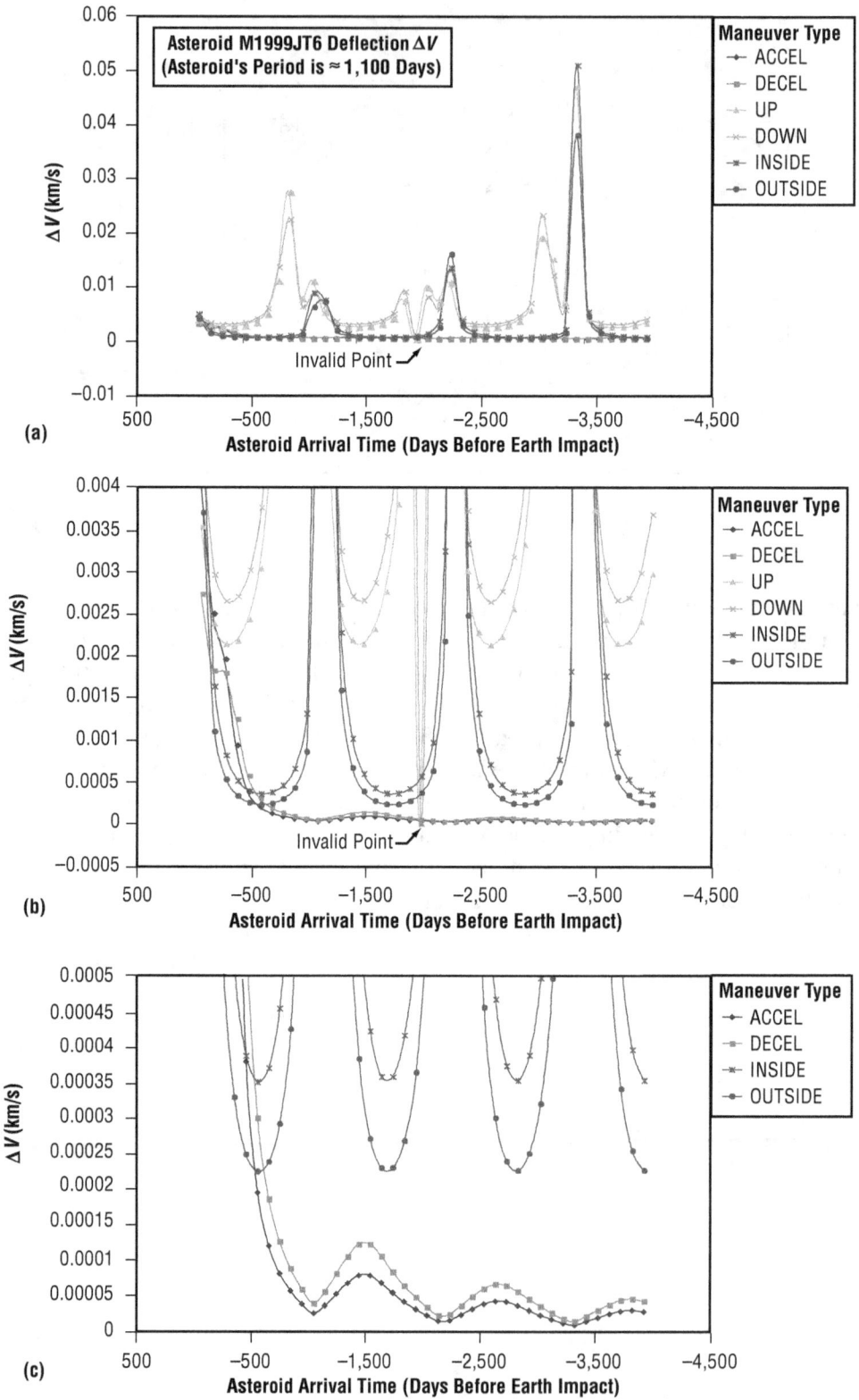

Figure 81. (a) Optimal deflection direction, (b) optimal deflection direction—detailed view, and (c) optimal deflection direction—detailed view—minimal deflection ΔV.

Once the deflection ΔV is known, the mass driver can be sized. The mass driver analysis program, MassDriver, is not set up to use a ΔV for sizing, so an iteration loop is introduced. MassDriverDVCalc is a script that converts the total impulse provided by the mass driver into a ΔV and then compares it with the ΔV required to deflect the asteroid. The optimizer, GoalSeek, changes the amount of asteroid mass ejected by the mass driver until the two ΔVs are equal. There is a simple unit conversion in the loop that converts ejected asteroid mass, or total expellant mass, from metric tons to kilograms. When the mass driver produces the correct ΔV, its total system mass value is passed to the staged chemical tool, where it becomes the payload mass for the outbound vehicle. One interesting feature about the mass driver is that its total system mass remains relatively constant regardless of the deflection ΔV required. This is because no expellant mass needs to be carried out with the mass driver; the asteroid provides it all. There is only one term in the mass driver sizing relationship that scales with ΔV—the Expellant Storage facility mass. This facility houses the expellant mass prior to ejection, and for this study, its mass is equal to 0.01 times the total amount of mass to be ejected. From figure 82 one can see that this change in component mass results in a minimal shift in the total system mass.

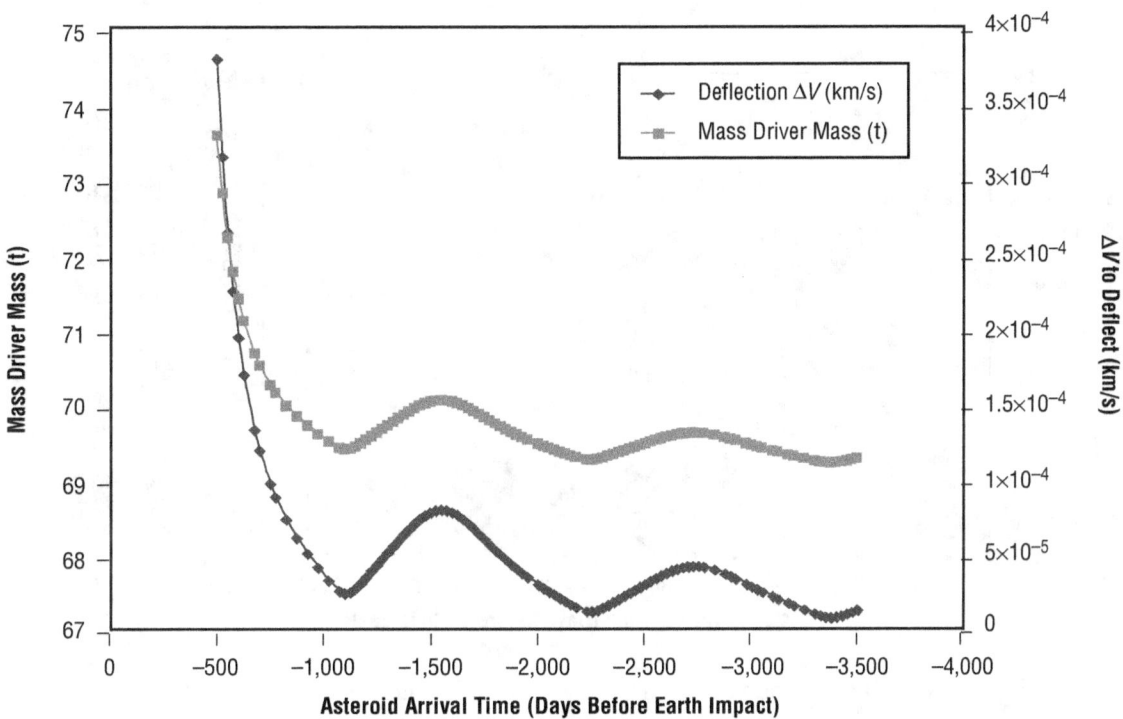

Figure 82. Variation of mass driver total system mass with required asteroid deflection ΔV for a 50-m-diameter chondrite.

The outbound trajectory code, Planetary Body—Outbound (PBO), is executed once the asteroid's position at rendezvous is passed from PBI. The outbound flight time is calculated and passed from the Main component. The required ΔV to rendezvous with the asteroid is output, and is passed to the staged chemical tool.

The staged chemical tool uses the payload mass given by the mass driver component and the ΔV from the outbound trajectory code, PBO, to size the vehicle that departs from Earth. For this scenario, only the lox/LH$_2$ propellant combination is considered, with an assumed vacuum I_{sp} of 465 s. Also, the vehicle was allowed five stages.

With all of the necessary components appropriately linked in a ModelCenter project file, the user can input a total mission duration and a rendezvous time, and then calculate the total mass required for the mission for a given asteroid composition and diameter. In this manner, the parametric analysis was performed for this mission scenario. The total mission duration, which is defined as the number of days between the vehicle's departure from Earth and the impending Earth-asteroid collision, was varied from 1,500 days to 10 yr, in increments of 25 days. The point at which the vehicle makes its rendezvous with the asteroid was varied from 500 to 3,500 days before impact, also in increments of 25 days. The small increment value was necessary to ensure that optimal orbital transfers were captured. Within these ranges, the outbound flight time could be as short as 25 days or as long as 3,100 days. Figures 83 and 84 show the resulting initial spacecraft mass at Earth departure.

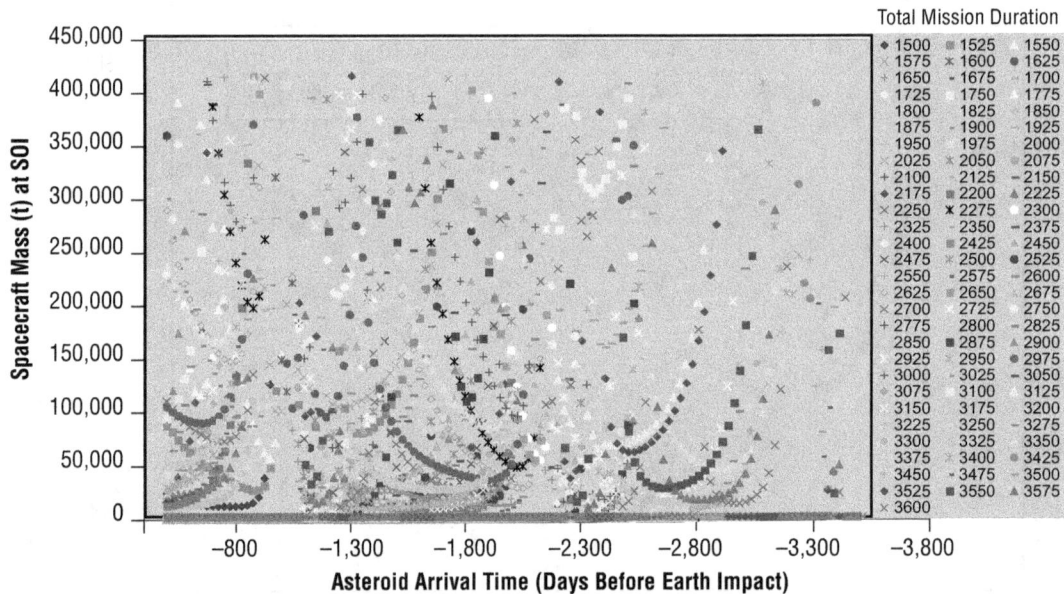

Figure 83. Staged chemical/mass driver vehicle mass at Earth departure.

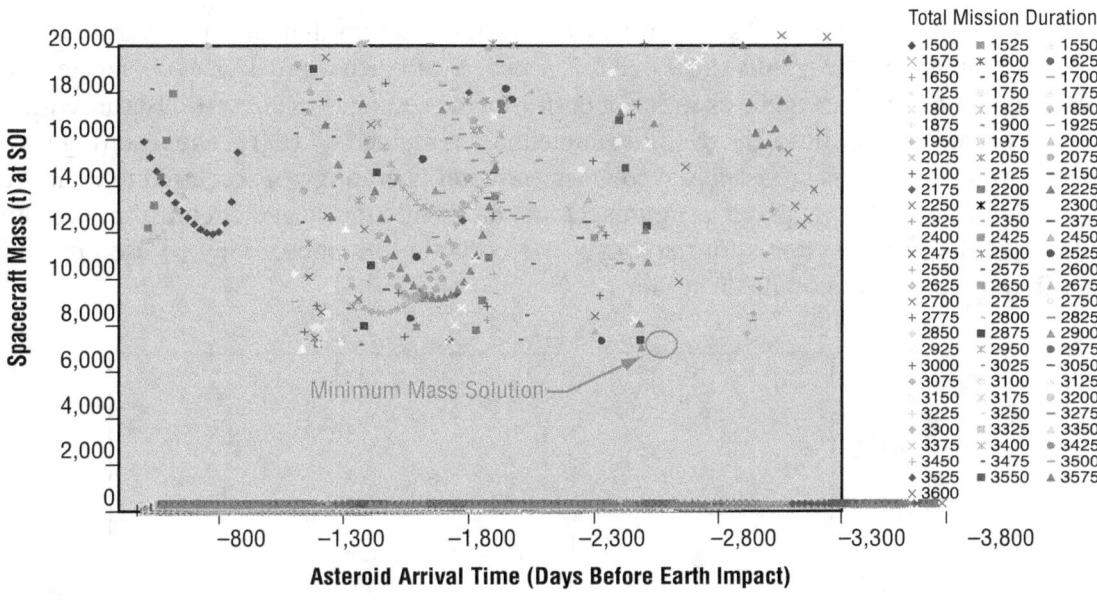

Figure 84. Staged chemical/mass driver vehicle mass at Earth departure (expanded view).

The minimum mass solution for this scenario is 6,850 t. This mission is launched 2,900 days before the asteroid's collision with the Earth, and arrives at the asteroid 500 days after launch. The trajectory is shown in figure 85.

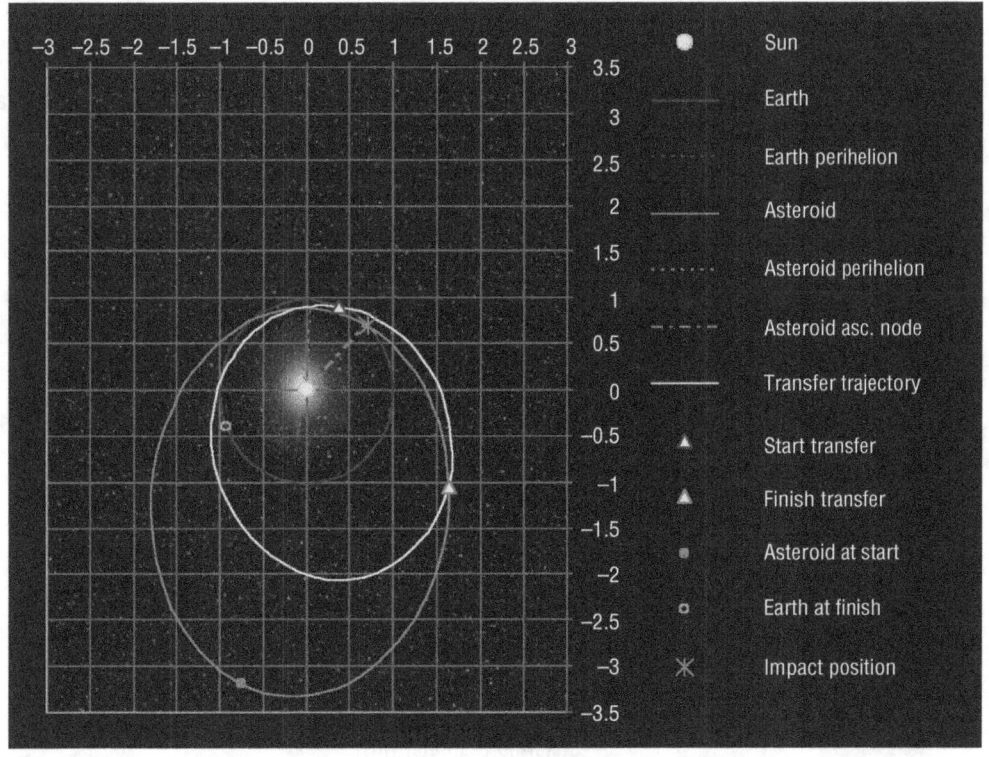

Figure 85. Optimal staged chemical/mass driver mission.

Once the optimal mission solution was found, the asteroid diameter was varied to determine its effect on mission mass. The results show that this scenario can be used to deflect chondrite asteroids up to 80 m in diameter before the staged chemical's performance becomes inadequate. Because the mass driver system mass changed very little, the overall vehicle mass remained <7,000 t. The resulting vehicle masses are given in figure 86. Mass driver deployed mass and total operating time required to deflect asteroids of 50 to 1,000 m in diameter are given in figures 87 and 88. While the staged chemical propulsion system is not capable of delivering systems for the larger asteroids, these numbers may be used in future analysis with a different outbound propulsion system.

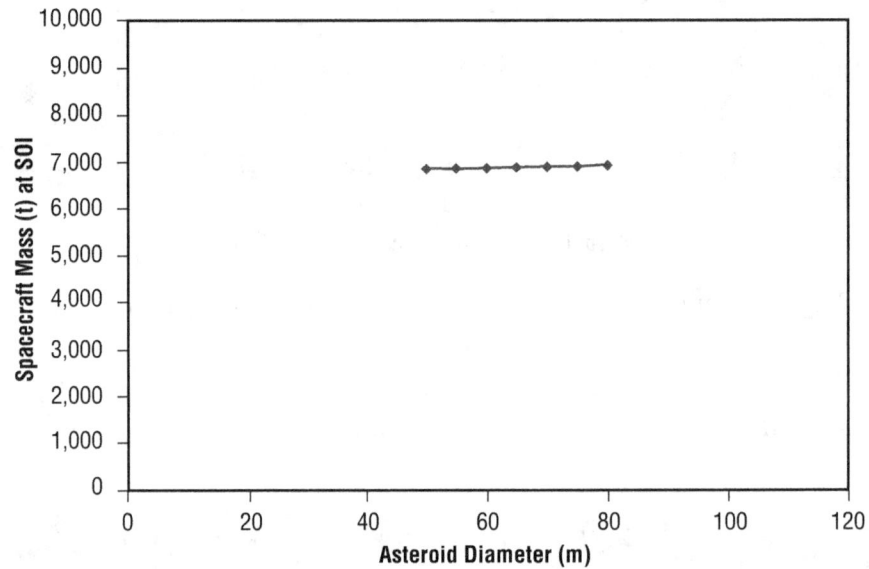

Figure 86. Staged chemical/mass driver vehicle mass versus chondrite asteroid diameter.

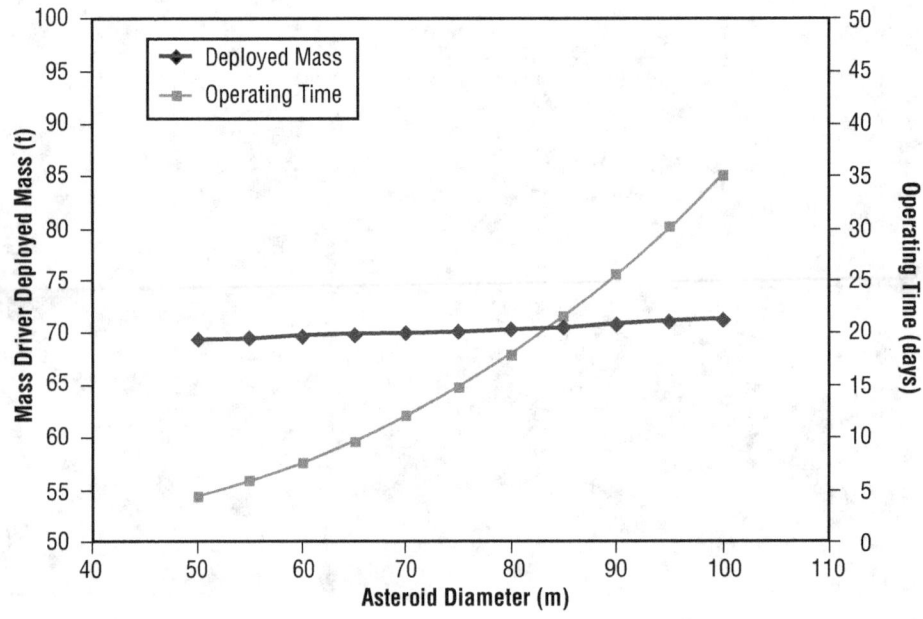

Figure 87. Mass driver deployed mass and total operating time versus chondrite asteroid diameter (50–100 m).

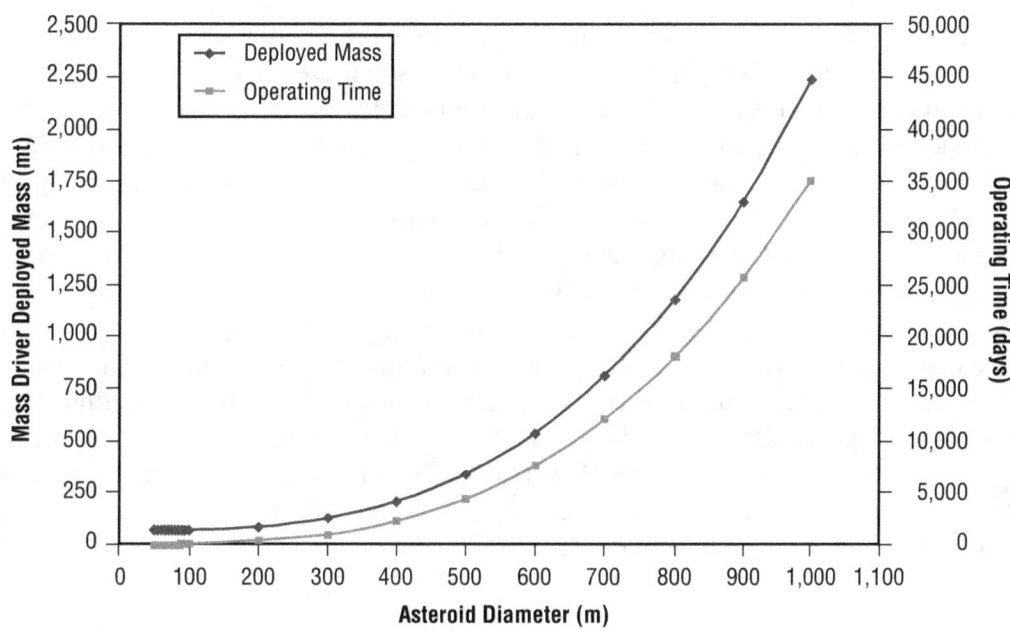

Figure 88. Mass driver deployed mass and total operating time versus chondrite asteroid diameter (100–1,000 m).

8.3.2 Staged Chemical/Nuclear Deflection

The cases considered in this scenario included both intercept and rendezvous with the asteroid. The overall layout is illustrated in figure 89. The main difference between this model and that for the staged chemical/mass driver option (which was considered in sec. 8.3.1) is the replacement of the mass driver with the nuclear deflection tool called explosion. Starting with the asteroid diameter and type—obtained from the main script—and the required asteroid ΔV from PBI, explosion calculates the size and number of nuclear blasts necessary to deflect the asteroid. The payload mass, which is just the total mass of the nuclear devices, is passed to the staged chemical tool, stagedChemical3. The staged chemical tool then sizes the chemical rocket that is required to deliver the nuclear devices from Earth's sphere of influence to the asteroid.

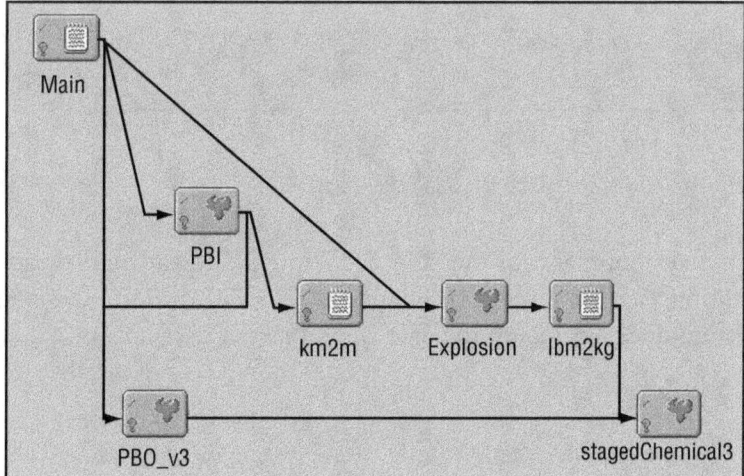

Figure 89. Diagram of the ModelCenter setup for the staged chemical/nuclear deflection option.

Various values of total mission time and rendezvous time are examined to determine those regions in which the global minimum total system mass resides. Results of these runs for the intercept and rendezvous cases are illustrated in figures 90 and 91. Once the regions of interest are identified, the Optimization component of ModelCenter is used to find the minimum total system mass. This is accomplished by iterating over both the total mission time and the rendezvous time. With so many peaks and valleys in the total system mass graphs in figures 90 and 91, it is necessary to restrict the search to specific regions before the optimization tool can be used. Otherwise, the iteration process is likely to converge on nonoptimal solutions. Through a thorough set of analysis runs, the minimum total system mass required to defeat a 100-m-diameter chondrite was found to be 847 kg for the intercept case (fig. 92) and 5,620 kg for the rendezvous case (fig. 93). For the intercept case, the optimum rendezvous and total mission times are 910 days before impact and 1,509 days, respectively. For the rendezvous case, the optimum rendezvous and total mission times are 132 days before impact and 1,075 days, respectively. These results are somewhat surprising, given the large staged chemical system that was required for the mass driver option, but they do illustrate the benefits of using nuclear energy to deflect the asteroid. The trajectories for the optimum solutions are shown in figures 94 and 95.

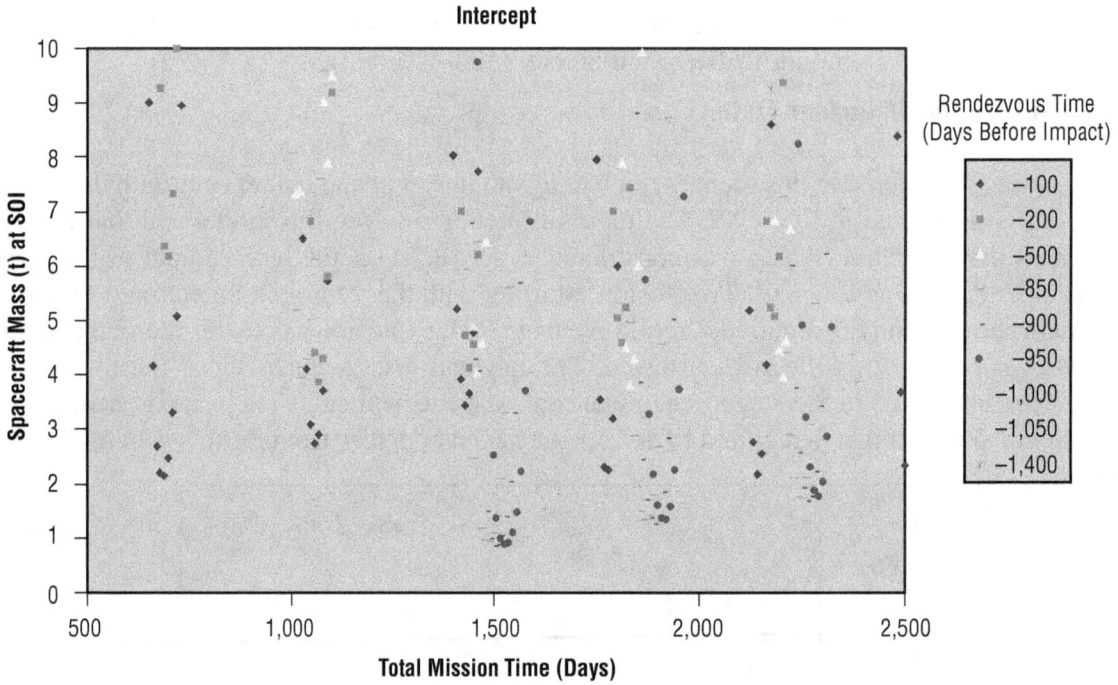

Figure 90. Total system mass for the staged chemical/nuclear blast option versus total mission time for various rendezvous times. Here, the staged chemical system does not match the asteroid's orbit at encounter.

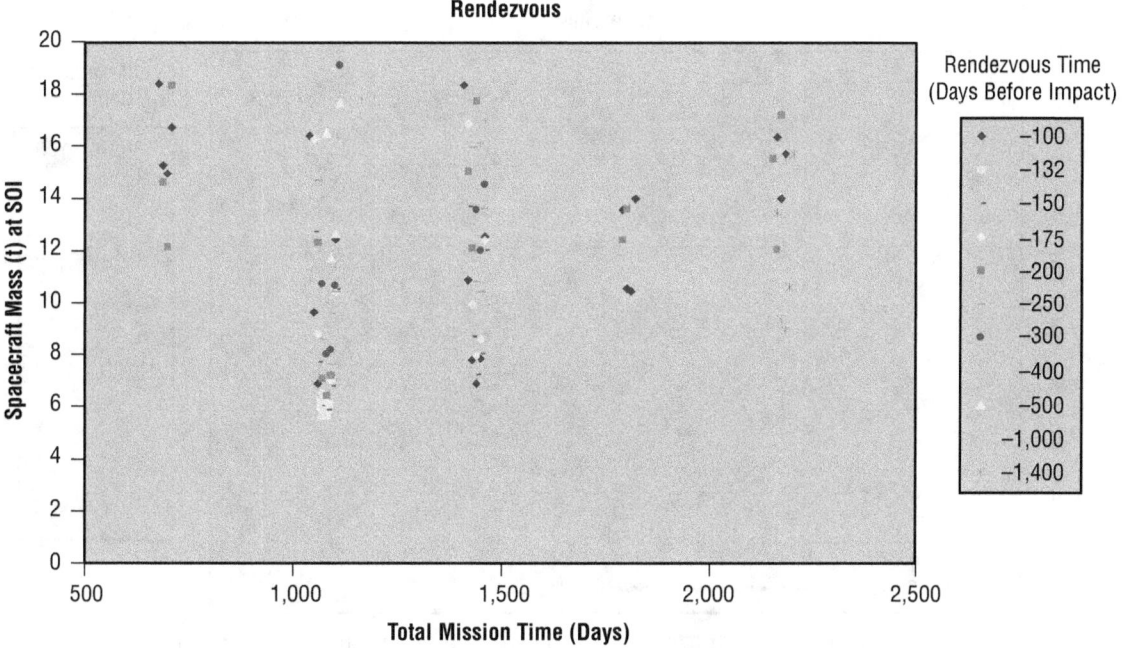

Figure 91. Total system mass for the staged chemical/nuclear blast option versus total mission time for various rendezvous times (zoomed). Here, the staged chemical system matches the asteroid's orbit at encounter.

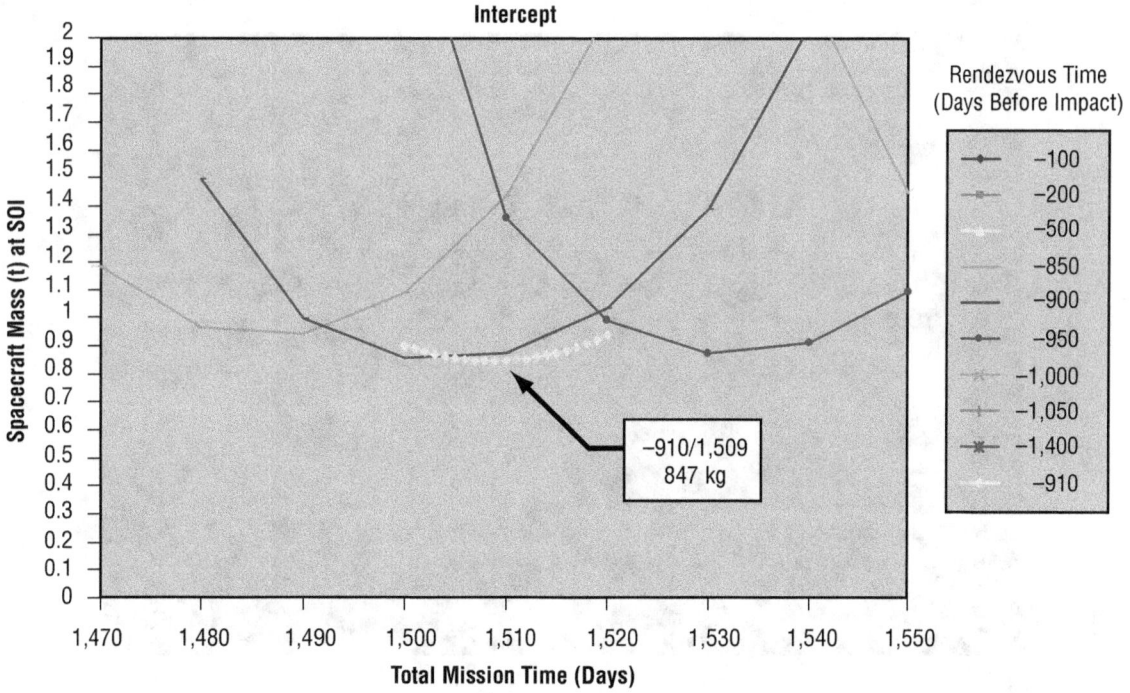

Figure 92. Minimum total system mass for the staged chemical/nuclear blast option, showing the optimum rendezvous and total mission times for intercept.

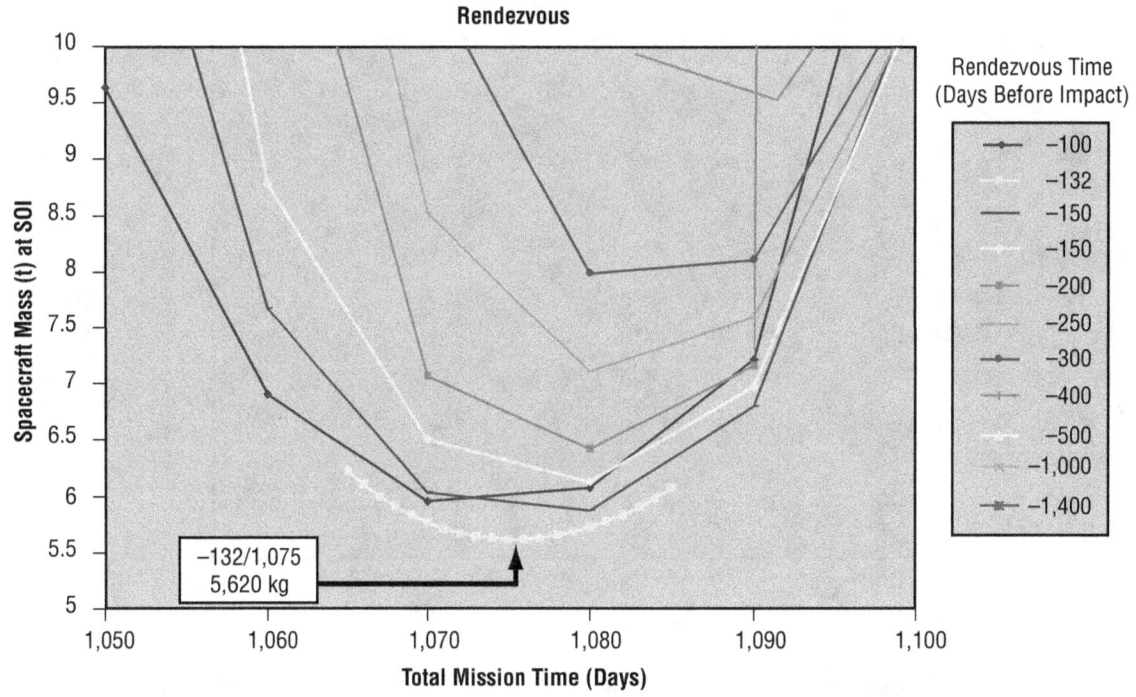

Figure 93. Minimum total system mass for the staged chemical/nuclear blast option, showing the optimum rendezvous and total mission times for rendezvous.

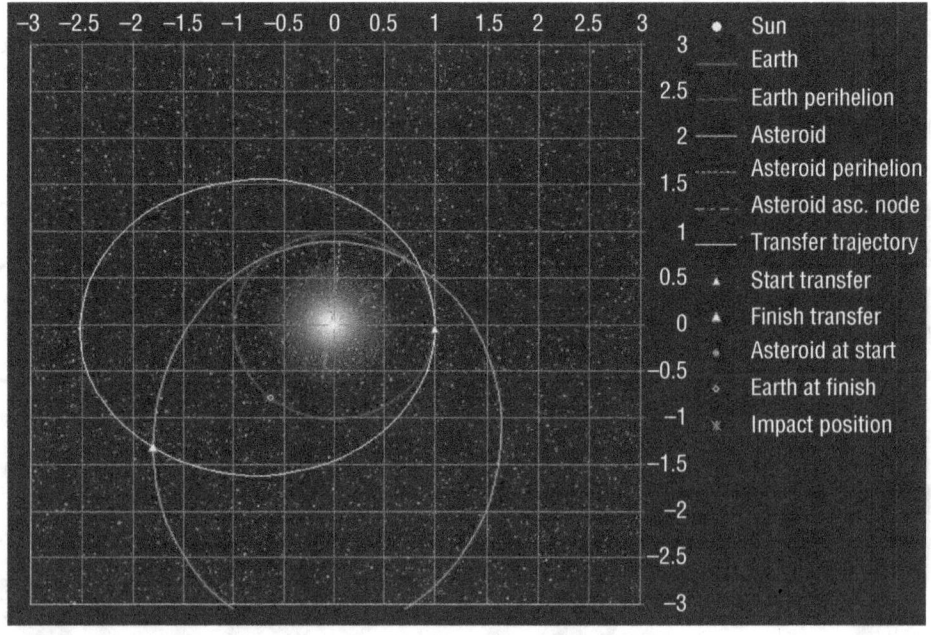

Figure 94. Optimum intercept trajectory for the staged chemical/nuclear deflection option.

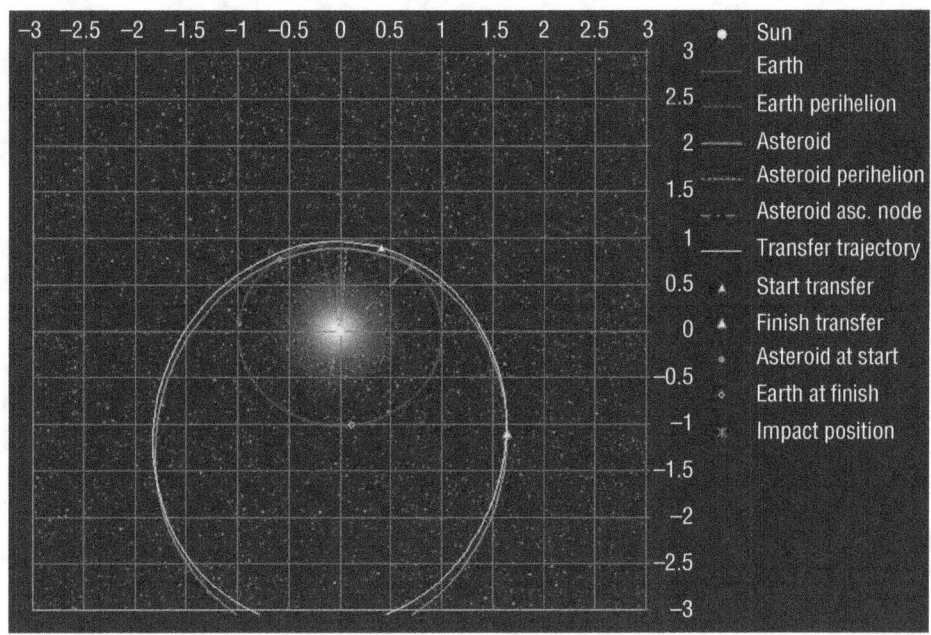

Figure 95. Optimum rendezvous trajectory for the staged chemical/nuclear deflection option.

The optimum rendezvous and total mission time vary little with changing asteroid mass; the optimum times are nearly the same for both 100- and 1,000-m-diameter chondrites. With this information, it is relatively easy to determine the required size of the staged chemical rocket versus asteroid size. The results are plotted in figures 96 and 97. With the rendezvous option, the total system mass increases rapidly and quickly exceeds the 1,000-t limit assumed for this study. However, total system mass for the intercept case is much smaller and is less sensitive to changing asteroid mass. Given the total system mass constraint of 1,000 t, the largest diameter M1999JT6 chondrite that this system can defeat has a diameter of 9,000 m for the intercept case and 1,000 m for the rendezvous case.

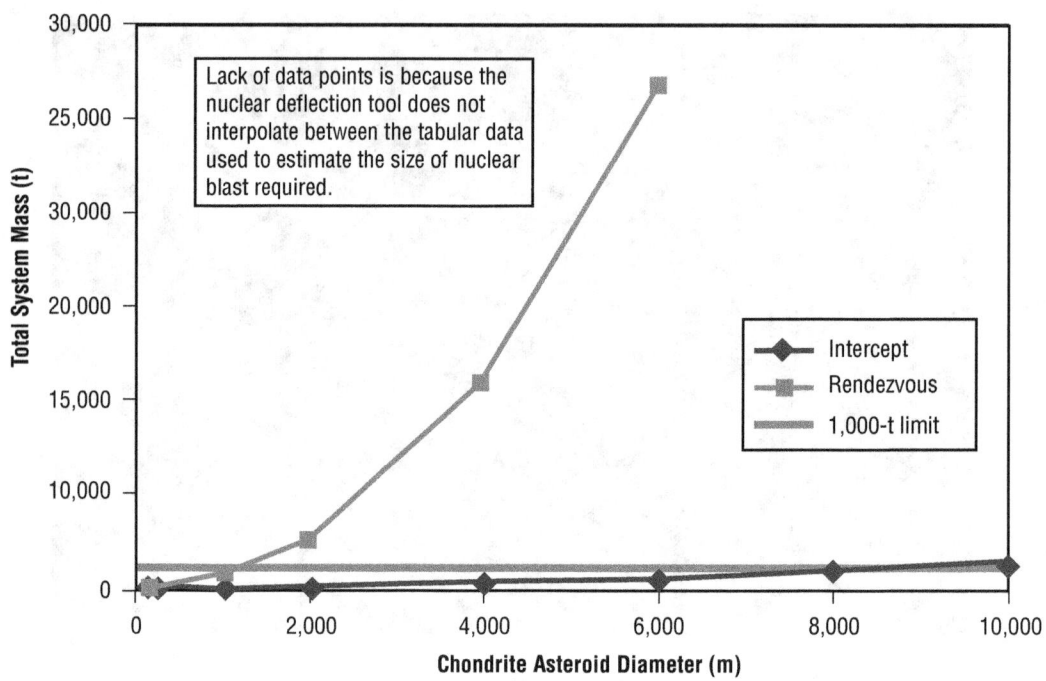

Figure 96. Minimum total system mass for the staged chemical/nuclear blast option versus chondrite diameter for both intercept and rendezvous.

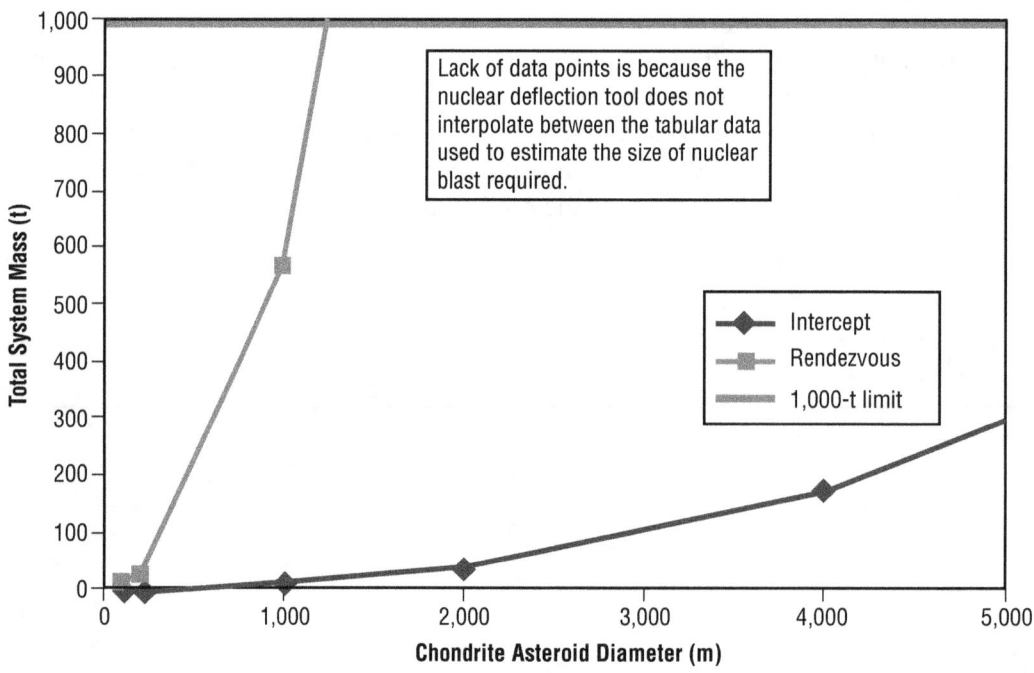

Figure 97. Minimum total system mass for the staged chemical/nuclear blast option versus chondrite diameter for the smaller chondrites, showing both intercept and rendezvous.

8.3.3 Staged Chemical/Kinetic Deflection

The staged chemical/kinetic deflection option consists of a chemical rocket that delivers a massive projectile to the asteroid. This projectile impacts the asteroid, nudging it off of its collision course with the Earth. The process flow for the analysis of this scenario in ModelCenter is given in figure 98.

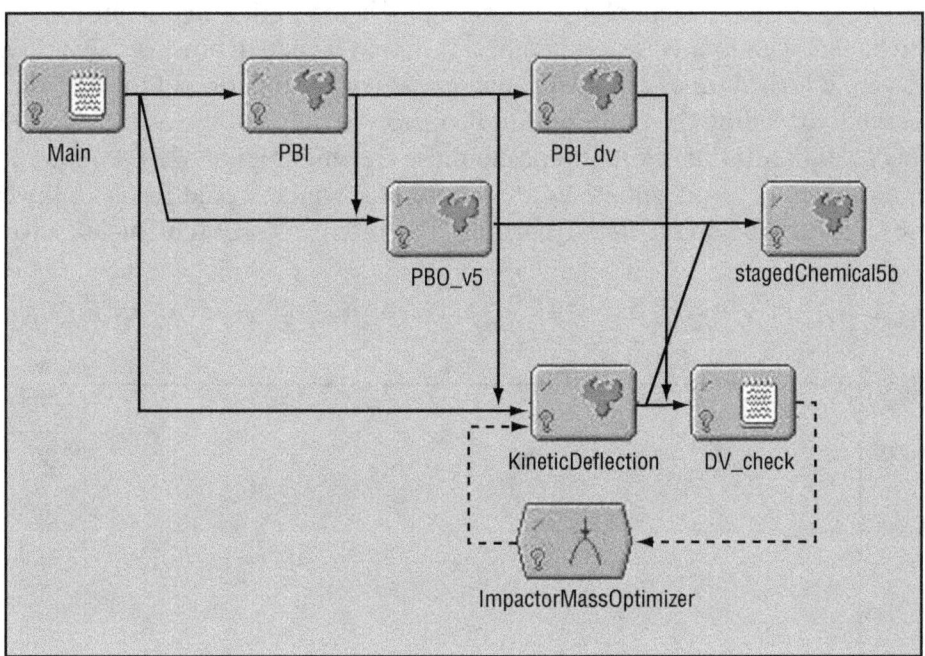

Figure 98. Staged chemical/kinetic deflection model.

The process is similar to that of the mass driver option, but there are a few exceptions. First, the diagram contains two instances of the inbound trajectory code, PBI and PBI_dv. The reason for this apparent duplication is that the asteroid's deflection direction is actually determined by the outbound trajectory of the projectile. Unlike the mass driver option, this vehicle impacts the asteroid; it does not rendezvous with it. Hence, the outbound trajectory, which determines the impact velocity, has a significant influence on the subsequent deflection. The two instances of PBI are required because the outbound trajectory code cannot run until the asteroid's position at interception is known. So, PBI runs first to determine the asteroid's position. This feeds into PBO, which gives PBI_dv the deflection direction. PBI_dv then produces the ΔV required to deflect the asteroid. Like the mass driver scenario, there is an iteration loop to ensure that the deflection ΔV produced is equal to that required. The projectile mass is varied until the momentum necessary to just achieve the desired ΔV is imparted to the asteroid. The resulting projectile mass is then input to the staged chemical tool as its payload mass. The staged chemical tool then provides the overall vehicle mass required to complete the mission. This time the number of stages is allowed to vary between one and five. All stages use the lox/LH$_2$ propellant combination, delivering an I_{sp} of 465 s.

The initial parametric study was performed for a 50-m-diameter asteroid of chondrite composition. Total mission time was varied from 150 to 1,200 days, and interception with the asteroid occurred between 100 and 1,125 days before impact. With a limited amount of time to complete this study, it was decided to limit the trade space for this case to nearer term missions only. The justification for this limitation comes from the fact that a simple kinetic deflection system, placed on an impact trajectory by a staged chemical propulsion system, is the simplest type of mitigation option that can be envisioned. It can be activated at relatively short notice—certainly much sooner than the more complex deflection options; e.g., the mass driver—and largely uses existing technology. In short, this is the option that is most readily available to counter near-term threats with limited reaction times. It should be noted, however, that even in this limited trade space, the asteroid arrival times considered encompass one full period of the asteroid ($\approx$1,100 days). One would expect that extending the trade-space to consider arriving at the asteroid 2,000 or 3,000 days before Earth impact would result in similar trends to those shown in figure 81. The required deflection ΔV should oscillate, diminishing in magnitude with each period, which would result in lower total mission masses than those presented here. See figure 99 for resulting vehicle masses for the missions considered.

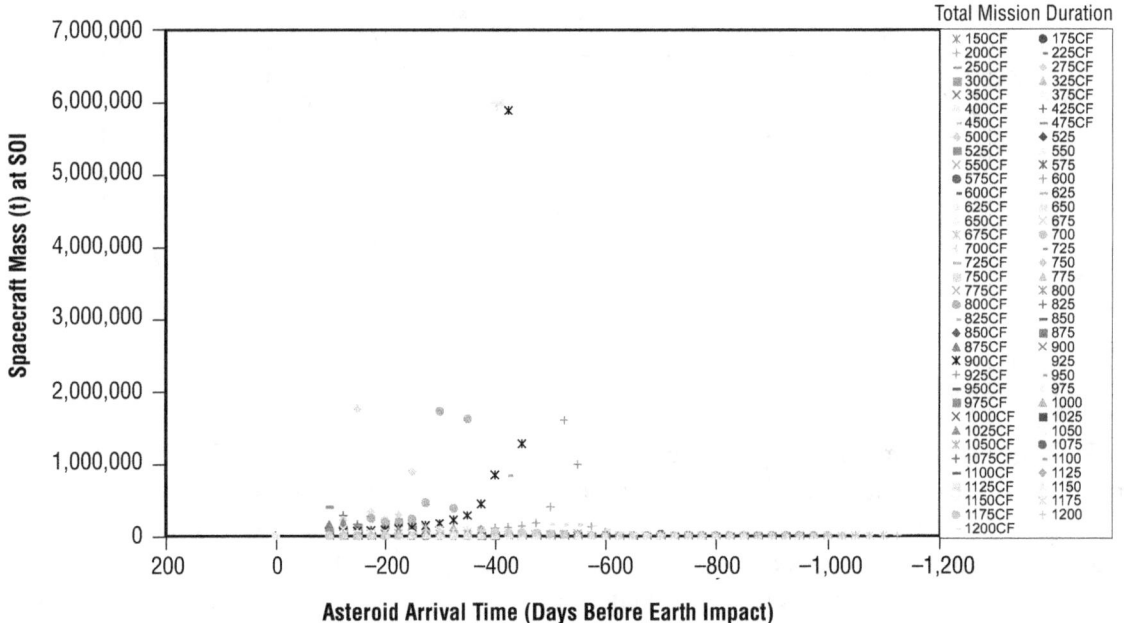

(a)

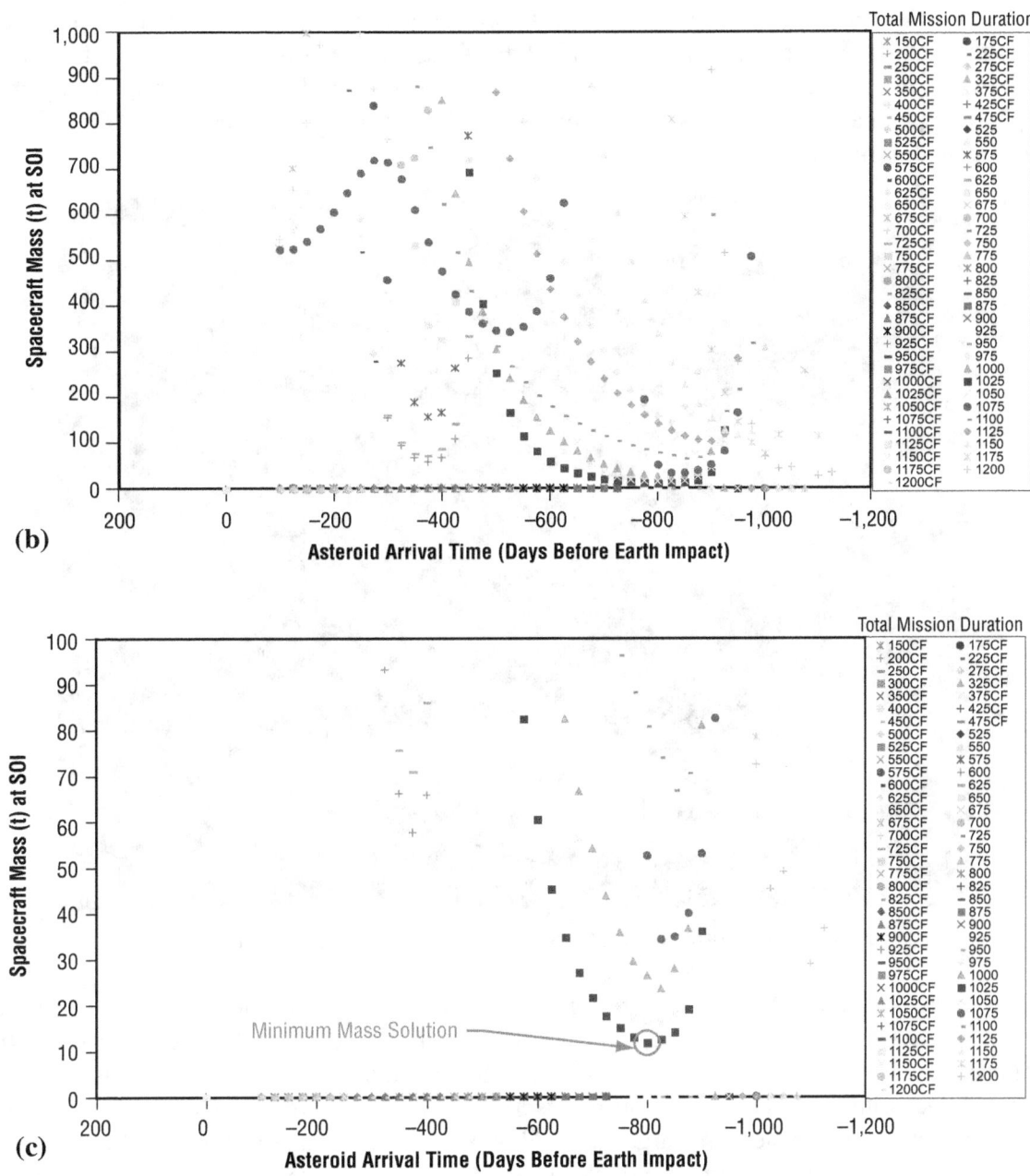

Figure 99. (a) Staged chemical/kinetic deflection vehicle mass at Earth departure, (b) detailed view, and (c) detailed view—minimum mass solution.

One additional complication associated with the kinetic deflection option is the possibility that the impact might cause the asteroid to fragment. Small impacts will produce only craters. Larger impacts will produce larger craters; however, there is some threshold impact size above that which the asteroid will actually break—a process referred to as catastrophic fragmentation. This was discussed in section 5.7. In figure 99(c), missions that result in catastrophic fragmentation are denoted by a CF after the series label.

The minimum mass mission is 11,853 kg. It corresponds to a 1,025-day total mission duration—defined as the time from Earth launch to possible Earth-asteroid collision—and impacts the asteroid at 800 days before Earth impact, or 225 days after the vehicle departs Earth. This mission causes only cratering, not complete fragmentation. It requires four chemical stages to provide the necessary ΔV for the outbound trajectory (plotted in fig. 100).

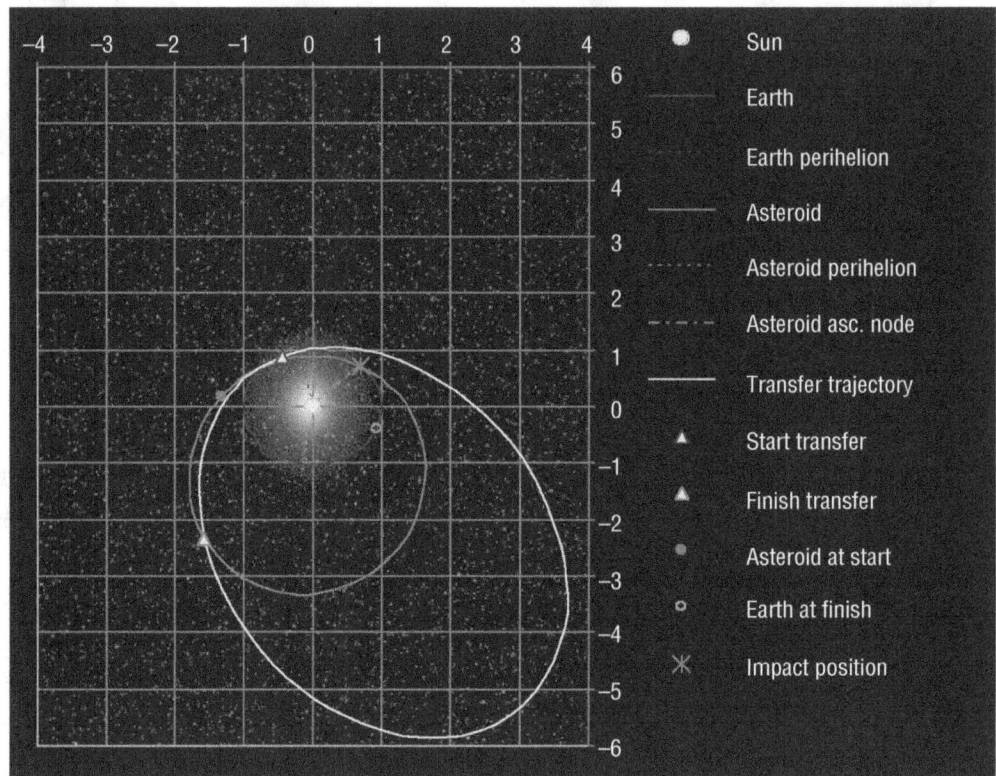

Figure 100. Optimal staged chemical/kinetic deflection mission.

The interceptor mass for this mission is 668 kg and the dry mass of the final lox/LH$_2$ stage is 259 kg. As the system is currently conceived, the final vehicle stage would jettison the projectile just before impact. The stage would then perform some small ΔV maneuver that would allow it to monitor the projectile's impact with the asteroid and transmit useful data back to Earth. Such data might be needed to plan future impacts on either the same target asteroid or some other similar body in the future. If the final stage were to remain attached to the projectile, it would serve to increase the energy and momentum available at impact. This might allow a reduction in the projectile mass, thus making the entire system smaller. This option has not been considered under the present study, but might be addressed in future work.

Once the optimal mission was determined, the asteroid size was again varied to determine its effect on the vehicle mass. As a result of this investigation, determination was made that chondrite asteroids up to 400 m in diameter can be successfully deflected; however, the resulting vehicle mass is very high for an asteroid of this size. Figure 101 shows the required spacecraft mass to deflect chondrite asteroids of up to 400 m in diameter. Figure 102 shows the projectile masses required to deflect asteroids of 50–1,000 m in diameter.

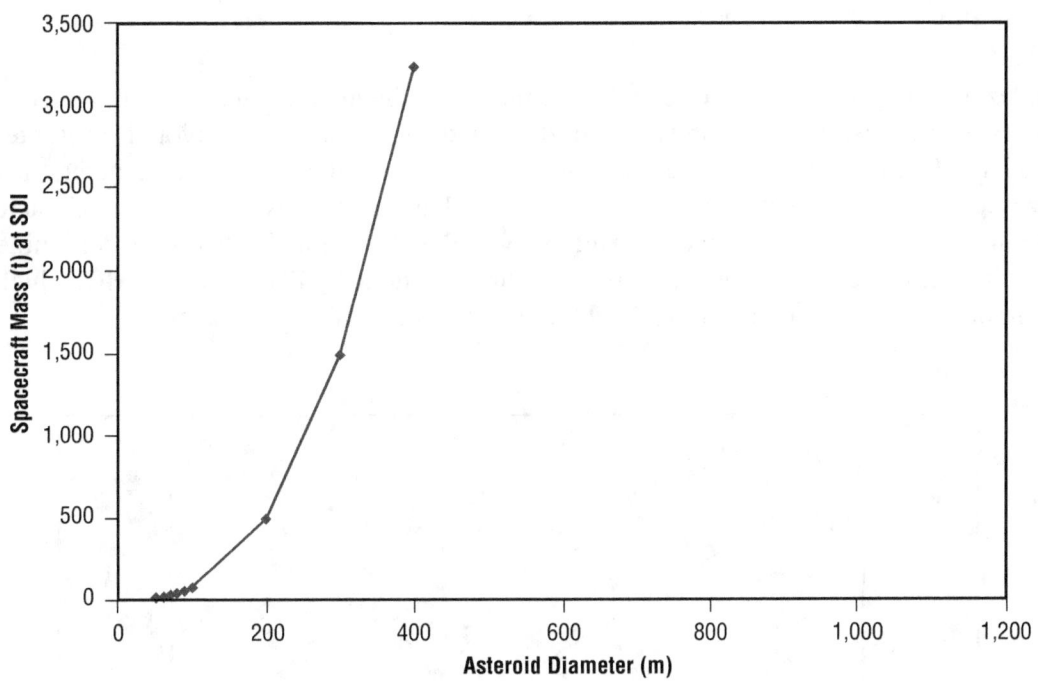

Figure 101. Staged chemical/kinetic deflection vehicle mass versus chondrite asteroid diameter.

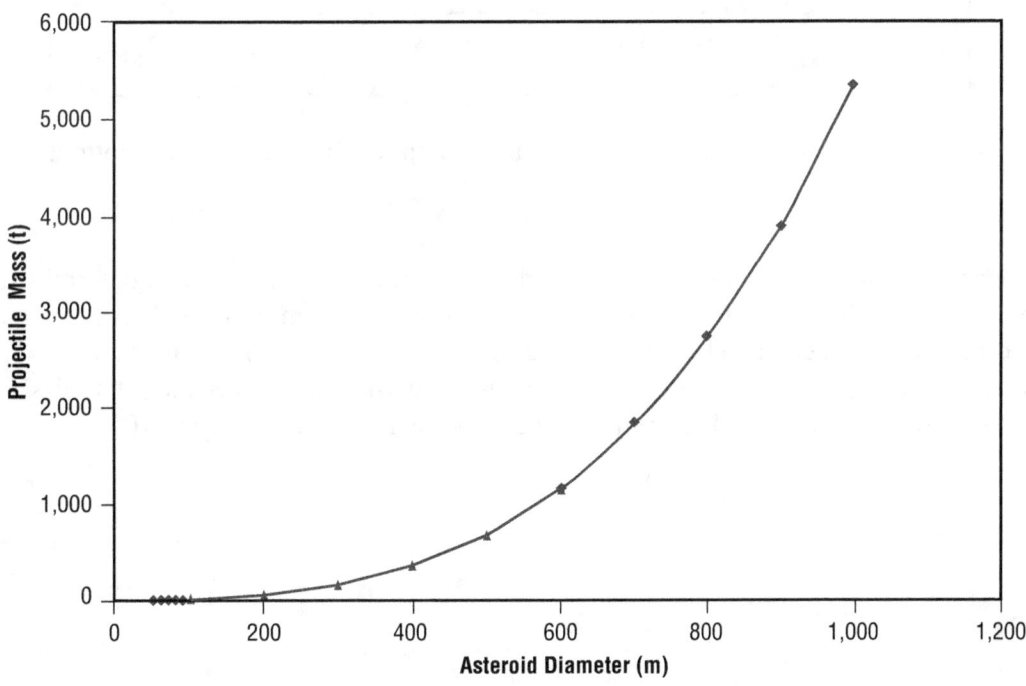

Figure 102. Projectile mass versus chondrite asteroid diameter.

8.3.4 Nuclear Pulse

Rendezvous with the asteroid or comet is required for the nuclear pulse option, which operates as both the outbound propulsion system and as the deflection device. The overall layout is presented in figure 103. Once again, the main script is merely a central location in which to input parameters; it passes the asteroid size and type to inboundNP, which is the tool that determines the number of nuclear pulses required to impart the required ΔV to the asteroid. As with the other models, this ΔV is determined by PBI. Sizing of the nuclear pulse system for the outbound journey is done by PBMExtPulseMC, which takes the mass from inboundNP; this mass is the payload that must be carried to the asteroid.

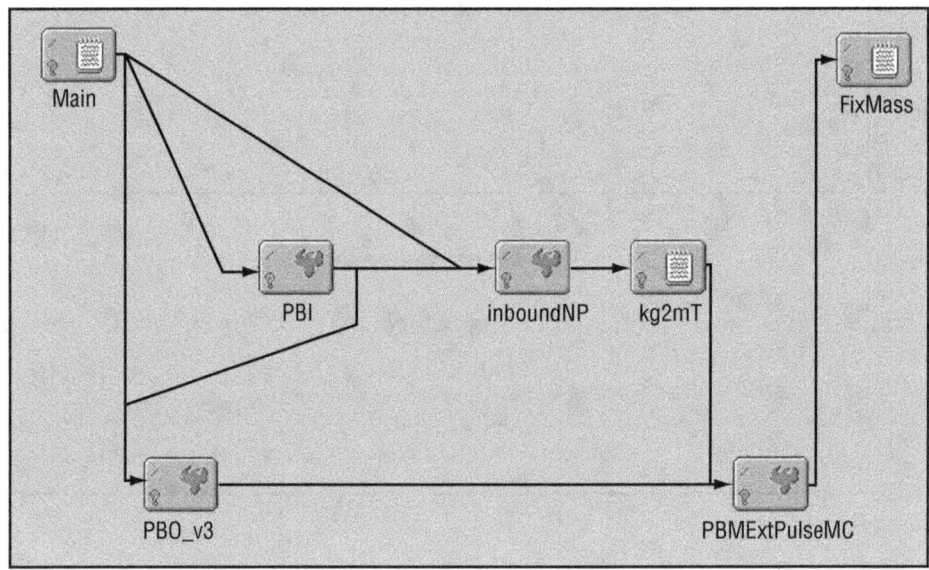

Figure 103. Diagram of the ModelCenter setup for the nuclear pulse option.

Various values of total mission time and rendezvous time were examined to determine the regions in which to concentrate the search for the global minimum total system mass. This is the same approach that was used for the staged chemical/nuclear deflection option (see fig. 90 for an example). The result of these runs is illustrated in figure 104. For the rendezvous case, the rendezvous and total mission times are 1,200 and 2,170 days, respectively. The optimum trajectory is illustrated in figure 105.

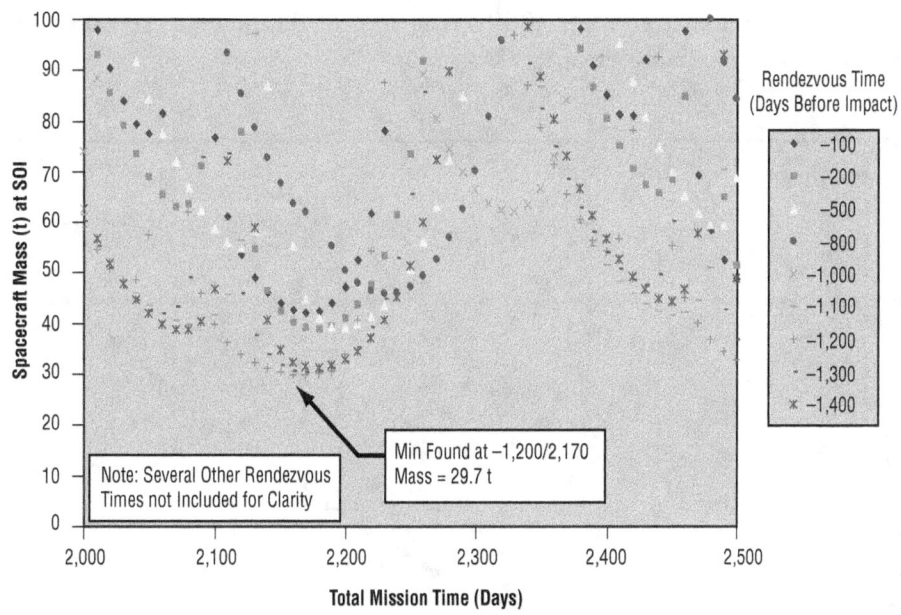

Figure 104. Minimum total system mass for the nuclear pulse option, showing the optimum rendezvous and total mission times.

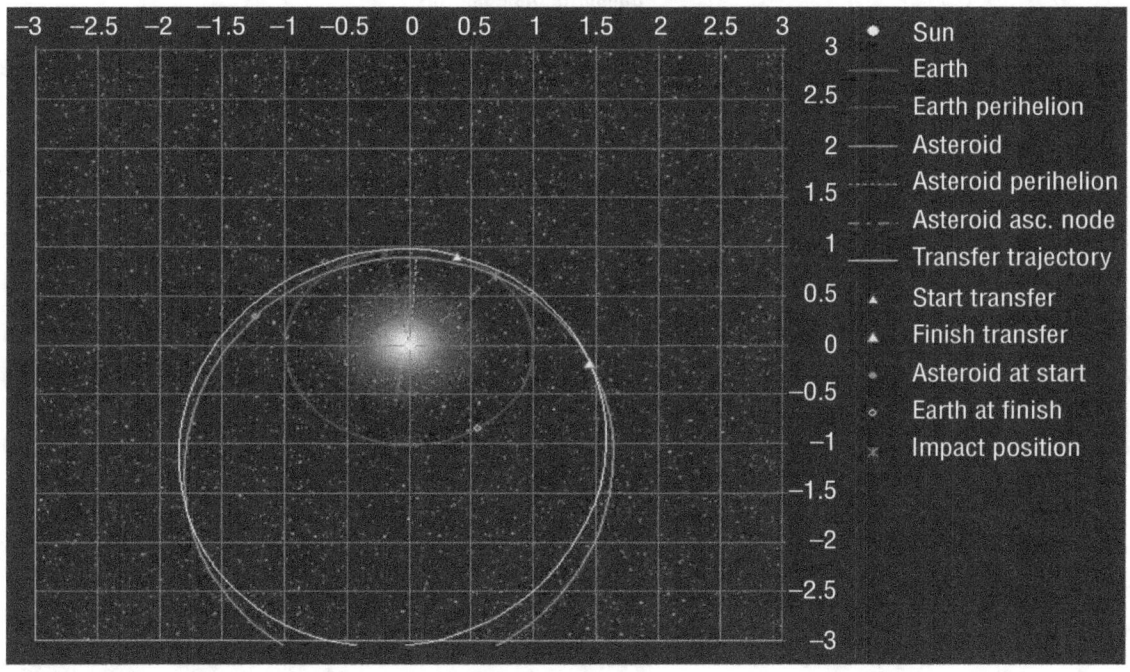

Figure 105. Optimum rendezvous trajectory for the nuclear pulse option.

As with the staged chemical/nuclear deflection option, the optimum rendezvous and total mission time varies little with changing asteroid mass. Therefore, it is relatively easy to determine the required size of the nuclear pulse system as a function of asteroid size. The results are plotted in figures 106 and 107.

Given the total system mass constraint of 1,000 t, the largest diameter M1999JT6 chondrite that this system can defeat would have a diameter of 9,000 m.

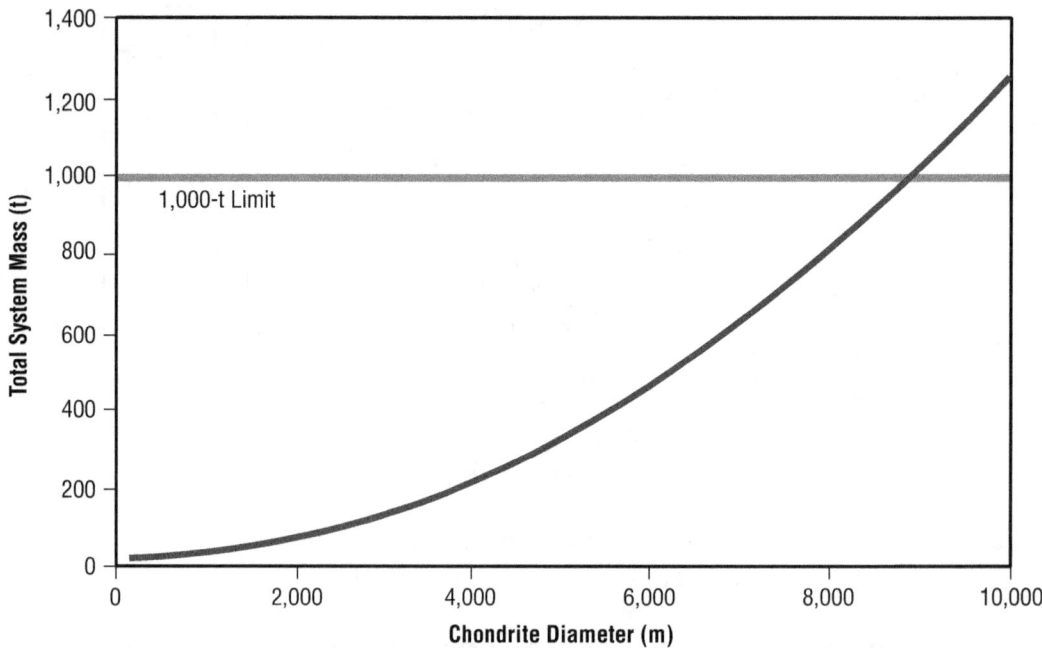

Figure 106. Minimum total system mass for the nuclear pulse option versus chondrite diameter.

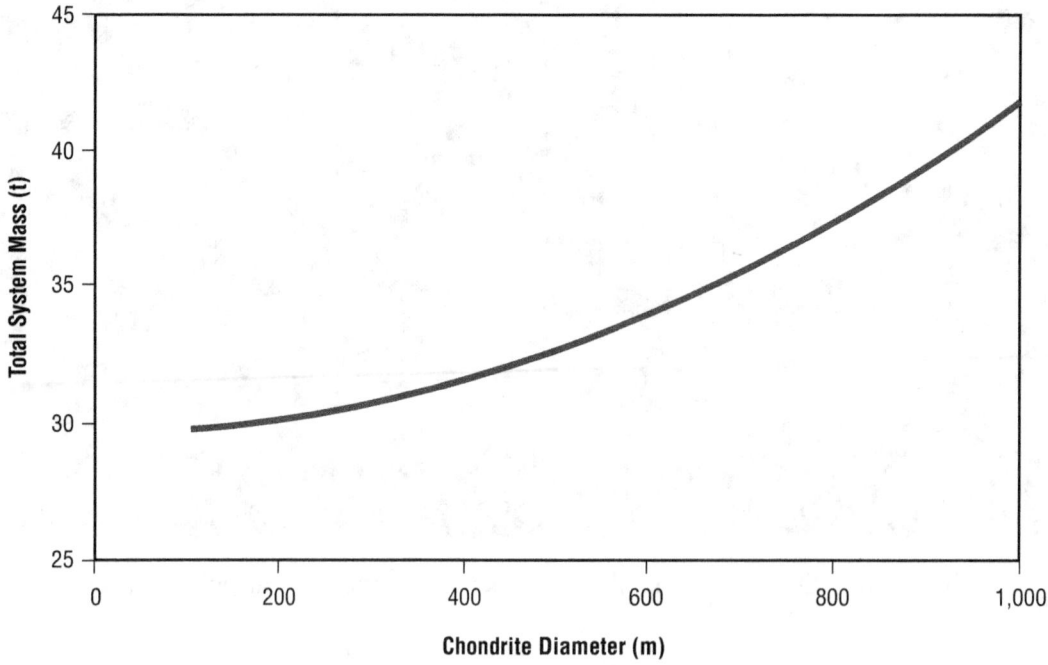

Figure 107. Minimum total system mass for the nuclear pulse option versus chondrite diameter for the smaller chondrites.

8.3.5 Solar Collector

Operation of the solar collector system was analyzed in a somewhat different manner than were the other systems. In this case, there is no payload that the outbound system must deliver, other than some avionics components, the RCS system, and some other minor components. For the solar collector, the inbound and outbound systems are the same. Also, unlike the other systems, the minimum total system mass continued to decrease as the total mission time increased. Therefore, rather than seeking a global minimum, the analysis instead located the minimum total system mass for two specific total mission times: 3 and 10 yr. In fact, these times were not exactly 3 and 10 yr, but were allowed to fluctuate by up to 100 days or so about the nominal values. This analysis method helps to illustrate the benefit of very long mission times with this system.

Picking an arbitrary rendezvous and total trip time usually results in either the outbound journey or the inbound journey dominating the solar collector size requirement; this does not yield an efficient solution. It seems logical that there must be some combination of rendezvous time and total trip time that would result in the collector being just large enough for both the outbound and inbound journeys. These points should yield the minimum total system mass, since the collector is optimally designed for both stages of the mission. Figure 108 shows the difference between inbound and outbound mass requirements for various total trip times. Negative values indicate that the inbound portion of the journey, which is the asteroid deflection portion, dominates the required solar collector size. For clarity, this plot does not include all of the total mission times that were examined, but it does show that for some values of total trip time, the inbound and outbound solar collector sizes can match at more than one rendezvous time. In such cases, however, the total system for one of the solutions was always significantly less than the others.

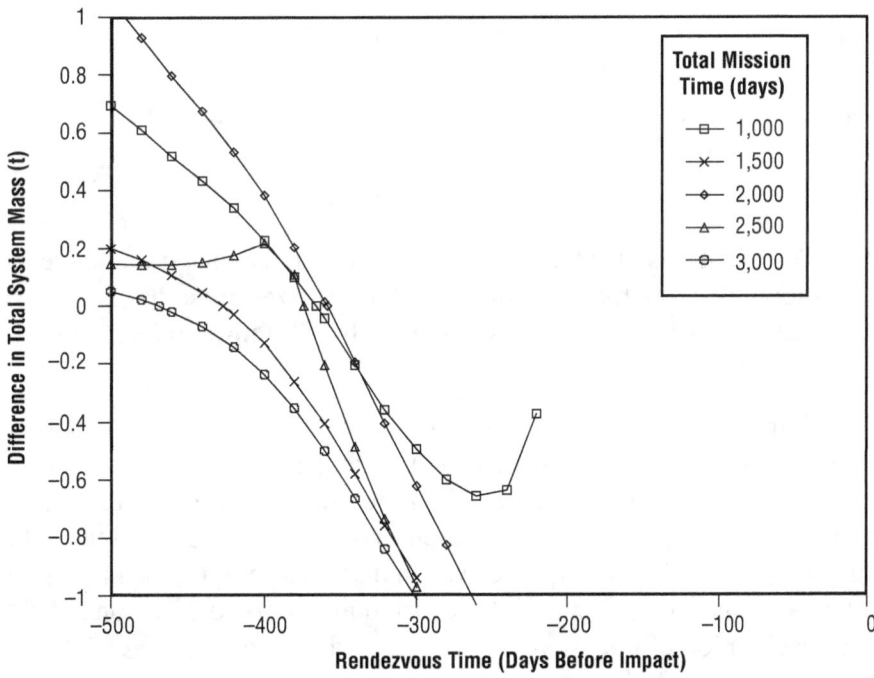

Figure 108. Plot of the difference between required outbound and inbound solar collector sizes. Negative values mean that the inbound requirement dominates.

The points at which the inbound and outbound size requirements are equal nearly always correspond to the points at which the total system mass is a minimum for a specific total mission time. In those instances where this is not the case, the total system mass is within a few kilograms of the minimum value. A plot of these locations for various total mission times, such as figure 109, reveals the cyclic nature of the total system mass; this is much like the other systems considered in this study. The plot also shows that the local minimum total system mass continues decreasing as the total mission time increases. Analyses far beyond the 10-yr limit for total mission time resulted in a continued decrease of the minimum total system mass. However, the objective of this analysis was not to find the global minimum total system mass, but rather to determine the minimum mass for two total mission times. As stated earlier, these two total mission times are around 10 and 3 yr—with some slight fluctuation to allow the ModelCenter optimizer to find the local minima.

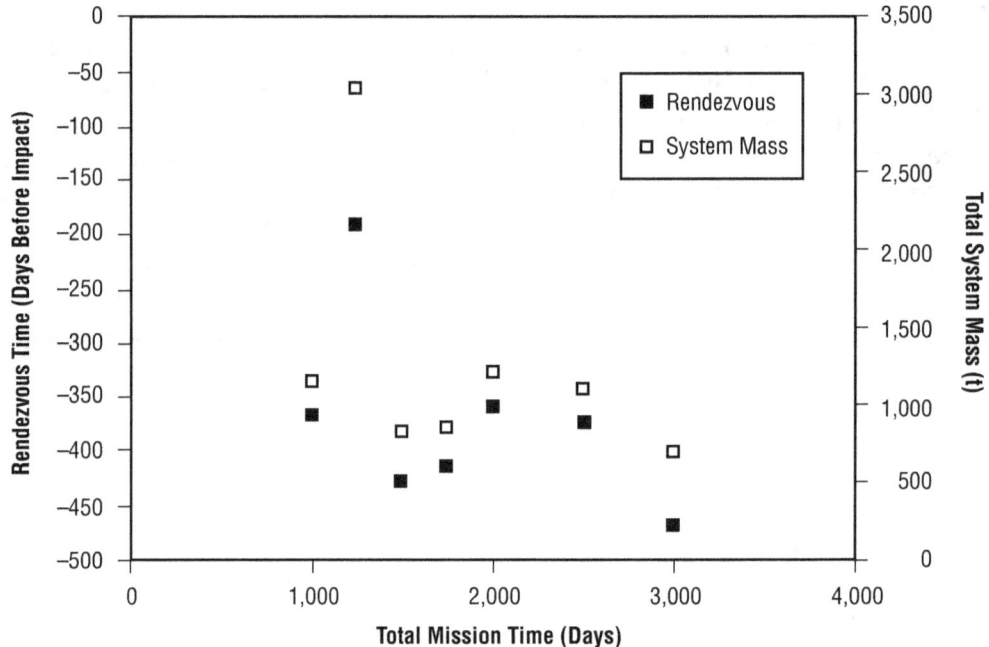

Figure 109. Combinations of total mission time and rendezvous time where inbound and outbound required solar sail sizes are equal, and the associated total system mass, for 500-m-diameter chondrite.

The relatively simple ModelCenter model is presented in figure 110. All tools in the figure have been described in previous sections, except for ssc, which is the solar collector sizing tool. As with the other models, PBI determines the asteroid deflection requirement, based on the rendezvous time, and PBO_v3 determines the outbound trajectory requirement based on the outbound time. The solar collector tool gives both the inbound and outbound required system masses as output; the minimum required system mass, which is the larger of the inbound and outbound requirements; dimensions of the solar collector; force on the asteroid; acceleration of the asteroid; and some additional data.

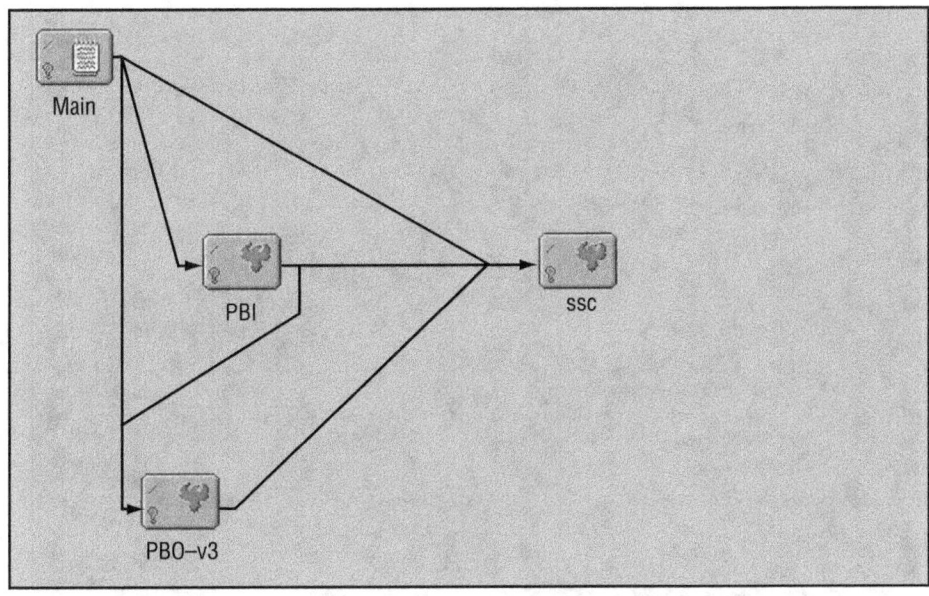

Figure 110. Diagram of the ModelCenter setup for the solar collector option.

Analysis of the solar collector option commenced near the end of the study. With limited time available for additional tool development, PBI and PBO_v3 were used to determine total ΔV requirements. Since PBI and PBO_v3 both determine impulsive ΔV requirements, the use of these values for analysis of the solar collector tool required the careful application of some conservative estimates. The force imparted on the asteroid by the solar collector was determined at a heliocentric distance of 1.5 au and assumed to remain constant. This constant value, coupled with the inbound flight time, determined the inbound ΔV requirement. This allowed the determination of the required solar collector size, provided the ΔV requirement for asteroid deflection was available. This analysis method does not accurately determine the required solar collector size, but it does highlight the trends in system performance for various total mission times, rendezvous times, and asteroid sizes.

For the staged chemical/nuclear pulse options, the optimum total mission and rendezvous times were very insensitive to asteroid size. This is not the case with the solar collector option: the optimum points vary with asteroid size and the rendezvous time changes considerably. For example, the minimum total system mass to deflect a 100-m-diameter chondrite occurs at a rendezvous time of 112.5 days before impact, with a total mission time of 3,636 days. But, the minimum for a 1,000-m-diameter chondrite occurs at a rendezvous time of 824 days before impact, with a total mission time of 3,711 days. Since inbound and outbound times are determined by asteroid mass, the size and type of an incoming asteroid would have to be determined quite accurately before the solar collector system was ever launched from Earth. Alternatively, a fairly large performance margin would have to be built into the system; i.e., a much larger than optimal solar collector would be required. For the rendezvous case, the optimum trajectory for the 100-m-diameter chondrite is illustrated in figure 111.

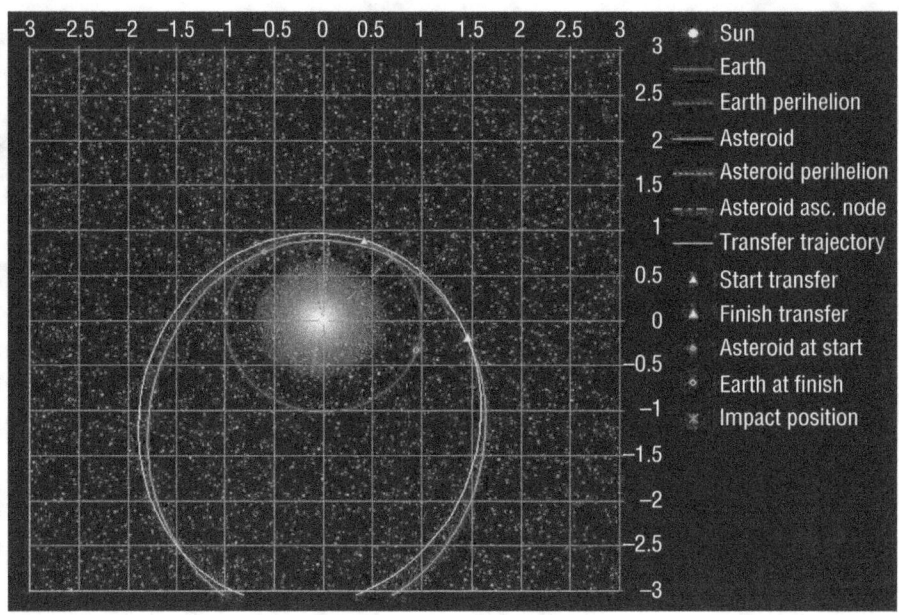

Figure 111. Optimum rendezvous trajectory for the solar collector option for a 100-m-diameter chondrite.

After extensive analysis, the required total system mass versus asteroid size for a chondrite was determined for two cases: total mission time of ≈3 yr and total mission time of ≈10 yr. These results are plotted in figures 112 and 113.

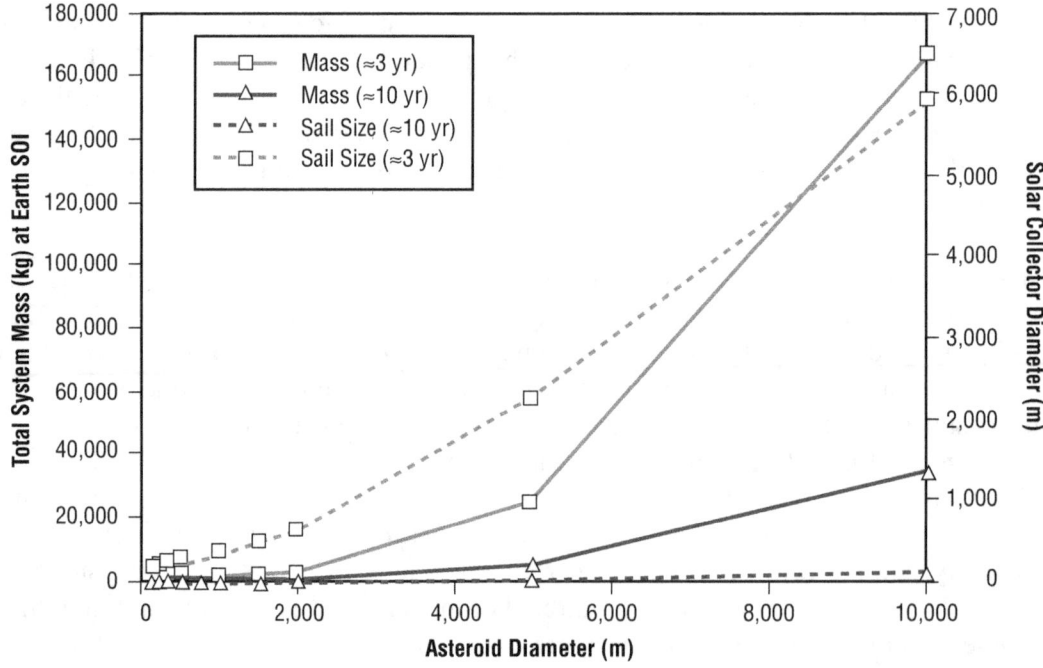

Figure 112. Minimum total system mass and size for the solar collector option versus chondrite diameter.

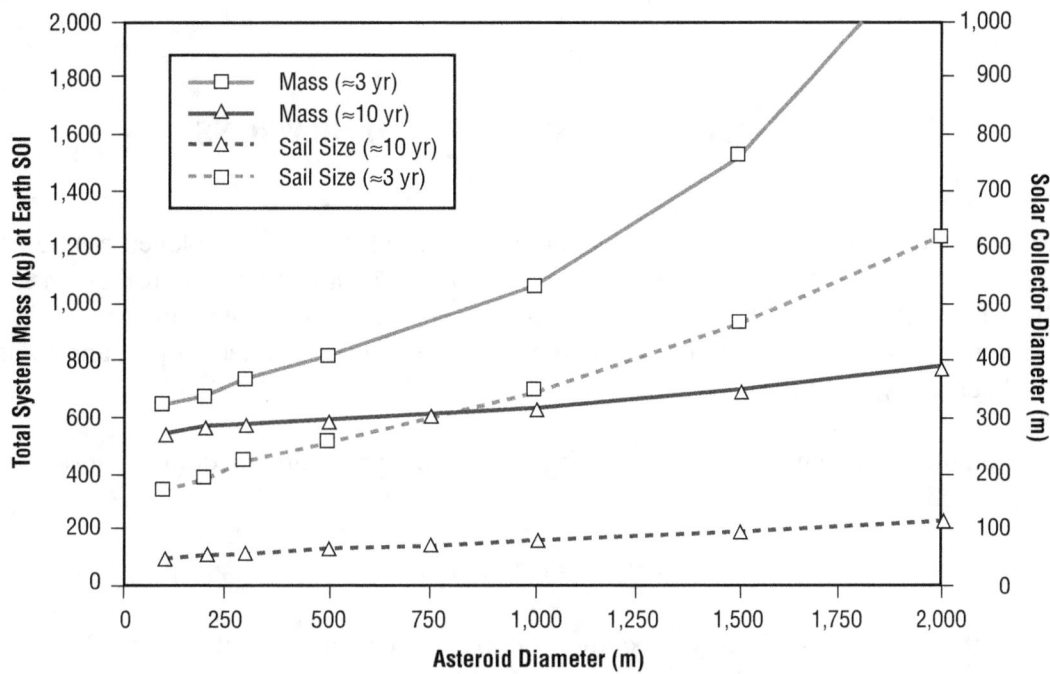

Figure 113. Minimum total system mass and size for the solar collector option versus chondrite diameter for the smaller chondrites.

Solar collector size, rather than total system mass, is the limiting factor for this option. Even for a 10-km-diameter chondrite, the total system mass is well within the 1,000-t limit imposed on the system mass. However, the diameter of the solar collector is a remarkable 6 km. Unfurling and controlling a collector of this size presents significant technical challenges. Despite these problems, it is clear that the solar collector could still be effective in deflecting an incoming asteroid.

9. CONCLUSIONS AND RECOMMENDATIONS

A wide range of potential mitigation techniques by which threatening planetary bodies could be either deflected or fragmented have been modeled in this study. Consideration was also given to a range of transportation methods by which the mitigation hardware could be moved out to an approaching body for either rendezvous or impact. Several possible combinations of mitigation techniques and transportation options have been analyzed in detail.

Although there is much that still needs to be done, conclusions and recommendations are given in sections 9.1 through 9.6.

9.1 Public Awareness

Despite the best efforts of Hollywood, the level of public awareness of this threat is still not high. Compared with other comparable threats, planetary body collision is still viewed as being a matter of science fiction rather than one of scientific fact.

While not advocating steps that could lead to hysteria and panic, the facts about this problem should be properly presented to the general public so as to raise public understanding of the threat and the ways in which it can be mitigated. Only when in full possession of the facts can the voting public make an informed decision about what steps should be taken.

9.2 Statistical Problem

The lack of attention given to this threat is in part due to a statistical problem. The chance of a significant-sized object striking the Earth is fairly low; such collisions might take place perhaps only once or twice per century. This has led to the danger being downgraded when compared with other threats to public safety, particularly those relating to acts of terrorism. However, the probability of an impact taking place cannot be considered in isolation; proper account must also be taken of the likely consequences of such a collision. Even the impact of a relatively small body would probably be very severe with fatalities in the millions, wide-scale destruction, and a recovery time possibly extending over decades.

To obtain proper assessment of the danger, some appropriate parameters, such as the expected number of fatalities over a period of, say, a decade or a century, must be considered. Table 29 shows the chances of death by a variety of causes for a typical resident of the United States. It is interesting to note that the probability of dying due to a planetary body impact is about the same as that of dying due to an aircraft crash. As table 29 shows, this approach presents an altogether more worrying perspective on the danger.

Table 29. Causes of death and associated probabilities for a U.S. resident.[45]

Cause of Death	Chance
Motor vehicle accident	1 in 100
Homicide	1 in 300
Fire	1 in 800
Firearms accident	1 in 2,500
Electrocution	1 in 5,000
Asteroid/comet impact	**1 in 20,000**
Passenger aircraft crash	1 in 20,000
Flood	1 in 30,000
Tornado	1 in 60,000
Venomous bite or sting	1 in 100,000
Fireworks accident	1 in 1,000,000
Food poisoning by botulism	1 in 3,000,000
Drinking water with EPA limit of tricholoethylene	1 in 10,000,000

9.3 Funding of Future Work

While a number of NEO search activities are currently underway, most are proceeding with very limited funding. In some countries, government agencies have declined to provide funding and, as recounted, actually scorned the detection efforts. A strong recommendation was made that funding for these efforts be increased. In particular, sufficient high-quality instruments must be made available to conduct an all-latitude observation program with the aim of cataloging the entire NEO population.

Although funding is limited for NEO surveys and searches, at least it is nonzero. By comparison, the study of mitigation techniques is—with the notable exception of this present effort—almost totally unfunded. Equally important, research into new mitigation techniques is nonexistent, except in those cases where the technology is under study for some other application.

A strong recommendation is made that a coherent study of mitigation techniques as well as their likely effectiveness, cost, and deployment times, be undertaken in the very near future. This study, which would represent an enlarged follow-on to this work, should involve and call upon all of NASA's considerable resources, as well as those of the DOD, the Armed Forces, and other government agencies; e.g., the National Oceanic and Atmospheric Agency, the Federal Emergency Management Agency, etc. International collaboration and funding should also be actively sought.

9.4 Development and Deployment of Mitigation Systems

The technical work undertaken in this study shows clearly that, although the mitigation challenge is formidable, it is not beyond our capabilities, provided preparations are begun well in advance of an impact. Despite the impression given by Hollywood, it is not practical to wait until a specific threat is identified before starting work on the mitigation system. Systems engineering, system deployment, and in some cases, technology development, will take several years.

Most strongly recommended is that, following an appropriate study phase, a development program be initiated immediately, with a view to deploying an operational system as soon as possible. It is already clear to us that a first-generation protection system will not be able to counter all possible threats; however, it should be able to defeat those most likely to occur. At the outset of Project Apollo, it was said that, while the United States could not guarantee to come first in the race to the Moon, failure to act would guarantee that she would come last. In the same way, it might be said that, while success in protecting the Earth against a cosmic impact cannot be guaranteed, failure to act will, in the long run, guarantee a major catastrophe of regional, if not global, proportions.

9.5 Accomplishments

It was not the intent of this study to select a particular technical option for recommendation as a threat mitigation system. Instead, our intent was to study the various options, in several cases, using improved and updated modeling techniques. It was also our aim to categorize these options into different mission configurations and to propose a method for comparing the large number of possible combinations of mitigation options and mission configurations. It was also our intent to recommend future work.

Several new tools were created during the course of this project. None should be regarded as a finished product and all would benefit from further development and refinement. As an example, the outbound trajectory tool is designed to give a first approximation of the required ΔV, using high thrust calculation methods. Similarly, the inbound tool takes a velocity vector at the point of impact and integrates the trajectory backward in time until the object is well outside the Earth's sphere of influence. It then integrates forward, after a deflection ΔV has been applied to the object, so as to determine the resulting miss distance from the Earth. The program iterates until a specified closest approach to the Earth has been achieved. Both these tools would benefit from the use of more accurate, although more calculationally intensive, techniques.

Numerous outbound propulsion systems and threat mitigation options were considered and modeled using several tools that were created by combining some basic principles of physics with engineering data available in the open literature. These tools yielded first approximations for the performance and mass of each technical option.

The data and tools available in the literature have been built on to create a threat assessment tool that calculates the percentage of the total threat that can be defeated using a given mitigation system.

A procedure for comparing all these technologies will be put into place in the future. In so doing, identification of mission categories for these technologies have been made and future analyses simplified by developing a procedure that deals with each category instead of attempting to deal with each individual technology combination separately.

9.6 Assessment of Mitigation Options

Although it was not the purpose of this study to select mitigation options, a preliminary assessment is possible. Table 30 summarizes the capability of each major system option.

Table 30. Summation of parametric results for mitigation concepts.

System	Maneuver	Time Before Impact (days)/Outbound Travel Time (days)	Total System Mass at SOI (t) for Different Asteroid Diameters (m)			Maximum Diameter of Asteroid*(m)/ Total System Mass at Earth SOI (t)
			100	1,000	10,000	
Staged chemical/ mass driver	Rendezvous	2,900/2,400	NA	NA	NA	50/6,849 80/6,918
Staged chemical/ nuclear deflection	Intercept	1,509/910	0.847	8.27	1,300	9,000/1,000
	Rendezvous	1,075/132	5.62	568	87,800	1,000/1,000
Staged chemical/ kinetic deflection	Intercept	1,025/800	73.8	NA	NA	260/1,000
Nuclear pulse	Rendezvous	2,170/1,200	29.7	41.8	1,240	9,000/1,000
Solar collector	Rendezvous (≈3 yr)	1,076/65**	0.637	1.07	167	§
	Rendezvous (≈10 yr)	3,635/115**	0.550	0.636	34.6	§

* Maximum was constrained to a total system mass at Earth SOI of 1,000 t.
** Times are for 100-m-diameter chondrite. Rendezvous times are greater for larger asteroids, although total missions times change little.
§ The solar collector system is limited more by solar collector size than by total system mass. See figure 112.

The mass driver was coupled with the staged chemical system to offer a non-nuclear threat mitigation option. However, the relatively massive mass driver system coupled with the least efficient stored-chemical system yielded unacceptably high initial masses. The mass driver could have been coupled with the solar sail/collector but would result in sail sides/diameters in the tens of kilometers. However, the mass driver is an attractive option for moving asteroids with the ultimate purpose of resource utilization. Thus, this concept should be carried forward in further studies, perhaps with effort expended to reduce the required mass driver system mass.

By comparison, the combination of a nuclear blast system and a staged chemical outbound propulsion system also offered excellent performance. Once again, it was the staged chemical propulsion system that limited the system performance.

A kinetic deflection vehicle, carried out by a staged chemical system, is theoretically capable of deflecting large asteroidal bodies. However, the interceptor vehicle mass required increases rapidly with asteroid size. Nonetheless, deflection of a 100-m-diameter asteroid is possible.

The nuclear pulse option performs well because of its use of the same, very effective technique for both outbound travel and deflection. Of all the options considered during this study, nuclear pulse offers the best prospect of providing an effective mitigation technique using existing and near-term technology.

The solar collector system showed itself to be capable, but only at the expense of a very large sail area and the consequent operational problems. As with the nuclear pulse option, it has the advantage of using a single unified system for both outbound propulsion and deflection.

Each of these options may well find some application in the future, but our initial results indicate that the nuclear pulse option offers the best defensive capabilities in the near term. This is by no means a recommendation but merely an observation based on the data at hand. Due to the level of fidelity and extensive assumptions that have been forced to be made in this limited study, it is recommend that all options discussed here, as well as other options suggested elsewhere, be carried forward into a higher fidelity analysis.

9.7 Future Work

A large amount of future work has been identified. All of our tools would benefit from more detailed analysis procedures. Many of the assumptions made during the development of our technology tools are in need of refinement. Our trajectory tools would benefit greatly from the ability to model continuous thrust propulsion systems. Our threat assessment tool requires more research into the available data on the asteroid and comet population. As a minor example of this, note that there are suggestions in the literature that cometary rings, such as the Leonid ring, may have nonuniform densities along their circumference. Since the Earth passes through such rings on a yearly basis, there would be a synodic period on which the Earth would cross these higher density areas, yielding a higher probability of impact. Our threat assessment tool also requires further research into the consequences of an impact.

There are several other mitigation options that were not studied because of resource and time limitations; two of these are of particular interest and merit some mention:

(1) Laser ablation is used as either a remote station or as a rendezvous option. This technique would allow deflection in a manner similar to that of the solar collector with a beam of high-energy coherent light being directed at the incoming asteroid or comet.

(2) This second new option involves firing inert masses from a mass driver located in Earth orbit, perhaps at a Lagrange point. This would combine our mass driver and kinetic deflection/fragmentation tools and would represent another remote station option.

Finally, a method to combine the quantitative results from this analysis needs to be established and the qualitative issues outlined for each technology in the outbound propulsion and threat mitigation sections in order to compare architectures.

After completion of the more advanced tools above, including the atmospheric reentry tool described earlier for fragmentation options, the overall threat assessment flow chart could begin, as described in figure 79.

9.8 Summary Conclusion

The threat posed by NEOs should be taken very seriously. It is well within humanity's ability to effectively defend itself against this threat. Development of the necessary technologies would also offer considerable synergy with NASA's other missions aimed at understanding the universe and exploring space. The planetary defense mission is also one for which NASA is uniquely suited and could potentially offer the Agency a goal that both fires the public imagination and creates a sense of urgency comparable to that during the Apollo program in the 1960's. The goal is to persuade those in positions of authority to continue the efforts presented here.

APPENDIX A—CURRENT NEAR-EARTH OBJECT SEARCH PROGRAMS

A.1 SpaceWatch

The SpaceWatch program is run by The University of Arizona's Lunar and Planetary Laboratory, established in 1980. Its primary goal is to explore the various populations of small objects in the solar system, and to study statistical data for asteroids and comets so as to understand the dynamical evolution of the solar system.

CCD systems scan the Centaur, Trojan, Main Belt, Trans-Neptunian, and Earth-approaching asteroid populations. The principal instruments, located on Kitt Peak, are the Steward Observatory 0.9-m SpaceWatch telescope and the SpaceWatch 1.8-m telescope. SpaceWatch is a pioneer in the use of CCDs and automation for asteroid and comet detection.

SpaceWatch currently has the distinction of having detected the smallest known asteroid— 1993 KA2, which is about 4 to 9 m in diameter—and has also observed the closest known approach of an asteroid to the Earth—1994 XM1, which approached to a distance of $\approx$105,000 km.

SpaceWatch continues to detect some 20 to 30 new NEAs per year.

A.2 Spaceguard

Spaceguard is an international association, established in 1996, to promote and coordinate activities for the discovery, monitoring, and orbital calculation of NEOs. It is intended to promote study activities at theoretical, observational, and experimental levels of the physical and mineralogical characteristics of the minor bodies of the solar system, with particular attention to NEOs. It is also intended to promote and coordinate a ground network—Spaceguard system—backed up by a satellite network for discovery, observation, and astrometric and physical studies.

The Spaceguard system is a collection of observatories engaged in NEO observations. There are currently more than 70 observatories registered, located worldwide. Wide ranges of instruments are in use.

Note that Spaceguard is a coordinating body and that the technology available and effort expended vary widely between the various participating observatories.

A.3 Lincoln Near-Earth Asteroid Research

Lincoln Near-Earth Asteroid Research (LINEAR) is a Massachusetts Institute of Technology Lincoln Laboratory program funded by the U.S. Air Force and NASA. Its goal is to demonstrate the application of technology, originally developed for the surveillance of Earth-orbiting satellites, to the problem of detecting and cataloging NEOs that threaten the Earth.

Equipment consists of a pair of 1-m-diameter, ground-based, electro-optical deep-space surveillance (GEODSS) telescopes at Lincoln Laboratory's Experimental Test Site on the White Sands Missile Range, Socorro, NM. The telescopes are equipped with Lincoln Laboratory-developed CCD electro-optical detectors and collected data are processed on site to generate observations.

Survey results as of April 2002 are as follows:

Number of observations to minor planet center	7,416,832
Number of asteroid detections	1,127,759
Number of new designations	157,920
Number of confirmed NEOs	951
Number of confirmed comets	82

A.4 Near-Earth Asteroid Tracking

Near-Earth asteroid tracking (NEAT) observatory is an autonomous celestial observatory developed by JPL and funded by NASA to study asteroids and comets. It is based upon a specially designed CCD camera.

The principal investigator is Dr. Eleanor F. Helin; co-investigators are Dr. Steven H. Pravdo and Dr. David Rabinowitz.

NEAT is comprised of two autonomous observing systems at the Maui Space Surveillance Site (MSSS)—NEAT/MSSS and at the Palomar Observatory—NEAT/Palomar. At both sites, the NEAT cameras use 1.2-m (48-in) telescopes to find NEOs, NEAs, and comets.

Nine new NEAs were discovered during July 2002: three Amors with one >1 km; five Apollos, including one PHA and one >1 km, and one Aten.

A.5 Lowell Observatory Near-Earth Object Search

The Lowell Observatory near-Earth object search (LONEOS) system can scan the entire sky every month, accessible from Flagstaff, AZ. It uses a 0.6-m Schmidt telescope and a CCD detector. It has been in operation since March 1998. The first new discovery was made on June 18, 1998, and is able to record objects to a magnitude limit near $V=19.3$, or $\approx 100,000$ times fainter than can be seen with the naked eye.

As of August 2001, LONEOS had submitted more than 1 million asteroid observations to the International Astronomical Union Minor Planet Center. It is estimated that, after 10 yr of full-time operation, LONEOS could discover 500 of the 1 km or larger NEOs and perhaps twice as many smaller NEOs, thus substantially increasing our knowledge of these bodies.

The asteroid discovery summary as of July 9, 2002, includes 10 Aten, 56 Apollo, 55 Amor, and 6 comets, for a total of 137 asteroids.

APPENDIX B—SOLAR ARRAY CALCULATIONS

The solar array forms an important part of the proposed mass driver system. The large distances anticipated between target asteroids and the Sun, coupled with the generally unfavorable incidence conditions, threaten to make the array one of the most massive system elements. Two distinct methods of sizing the array have been identified.

B.1 Method I

The electrical power (P) can be written as

$$P = FA\varepsilon ,\qquad(89)$$

where

F = solar flux (W/m^2)
A = array area (m^2)
ε = efficiency.

If σ = mass per unit area, then

$$A = \frac{M}{\sigma} ,\qquad(90)$$

where M = array mass.

Thus,

$$\sigma = \frac{FM\varepsilon}{P} ,\qquad(91)$$

and hence,

$$\sigma = \frac{F\varepsilon}{\left(\dfrac{P}{M}\right)} .\qquad(92)$$

For Earth orbit, $F = 1{,}300$ W/m^2. For current generation arrays, $\varepsilon = 0.15$ is achievable, and can be bettered. Reference 46 (p. 333) gives $(P/M) = 14$ to 47 W/kg. This implies $\sigma = 4$ to 14.

B.2 Method II[47]

Achievable specific power = 130 W/kg (assumed achievable at Earth with ideal array orientation with respect to the Sun). Let R_E = Earth's orbital radius and R_A = asteroid orbital radius.

Hence, achievable specific power at asteroid = $130 \times (R_E/R_A)^2$ W/kg. Again, it is assumed that the array is ideally oriented with respect to the Sun.

Introducing an additional degradation factor (α) to take account of (1) nonideal orientation, (2) possible asteroid rotation, (3) dust obscuration of array, ..., gives

$$\text{Specific power} = 130\alpha \left(\frac{R_E}{R_A}\right)^2 \left[\frac{\text{W}}{\text{kg}}\right]. \qquad (93)$$

APPENDIX C—MASS DRIVER

Appendix C contains details of a simple model for the key components of the mass driver system.

C.1 Model of the Forces on a Bucket Coil Due to the Nearby Drive Coils

As explained in section 5.7, although a bucket coil receives an accelerating force from the closest pair of drive coils, it experiences alternately retarding and attractive forces from each more distant pair (fig. 114). Of course, the more distant pairs of coils produce lower forces than do the nearer coils. For the force calculations in this analysis, only the effect of the first two pairs of coils will be considered. The nearest pair provides the major motive force; the next pair provides the major retarding force. By limiting consideration to these four coils, one essentially conducts a conservative analysis. This is because all subsequent drive coils—in theory, stretching out to infinity in both directions—can also be grouped into sets of four coils; the nearest and next-nearest sets are shown below.

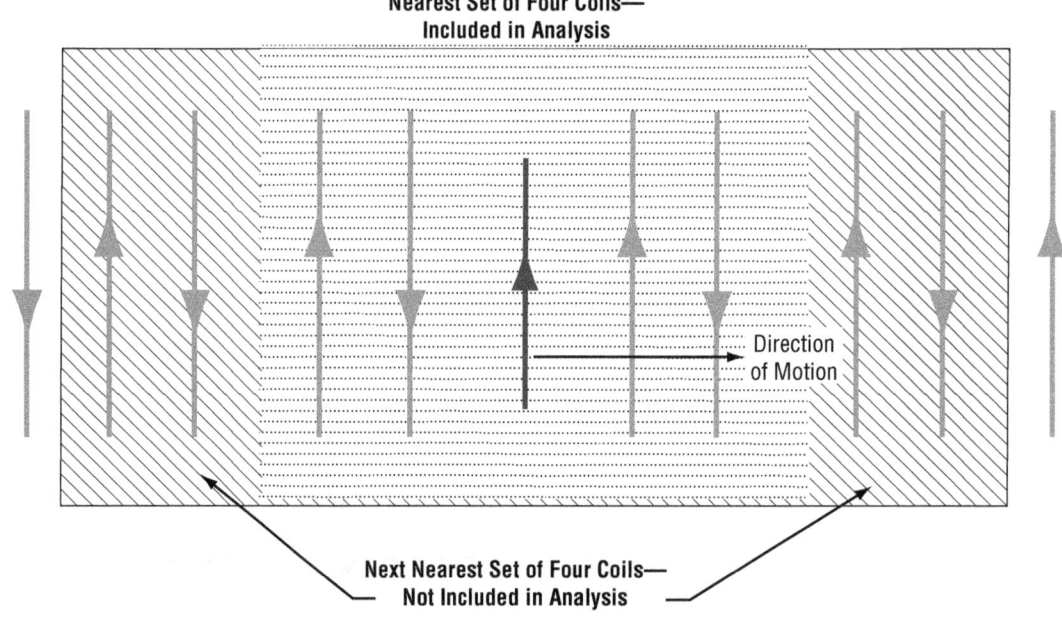

Figure 114. Drive coils included in and omitted from the analysis.

If one examines the coils which comprise the next-nearest set of four; i.e., the first coils to be excluded in the analysis, it is immediately apparent that the nearer pair exerts a motive force and the slightly more distant pair exerts a retarding force on the bucket coil. Hence, the net force from this set of four coils will be a motive one. The same argument can be applied to all subsequent; i.e., more distant, sets of four coils. Each produces a net motive force on the bucket coil, although declining in magnitude as distance increases. Hence, by their exclusion, the analysis neglects a portion of the overall motive force.

The analysis is conducted by considering the motion of a single bucket coil between two adjacent drive coils. The results can be multiplied to also include the effect of the remaining three-bucket coils. Figure 115 shows the bucket coil and surrounding four drive coils.

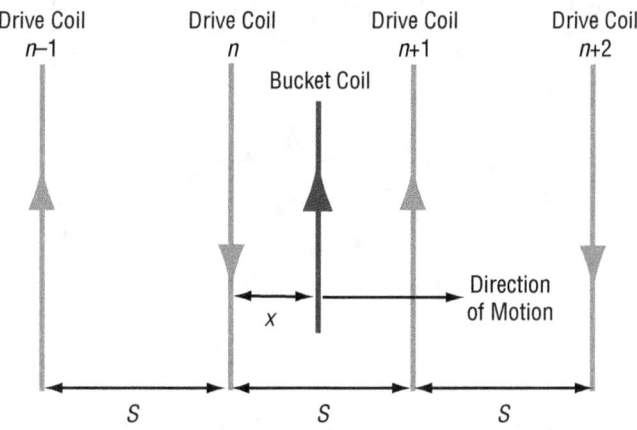

Figure 115. Drive coils included in the analysis.

The force (F) between any two coils (designated as 1 and 2) is given by

$$F = N_1 N_2 I_1 I_2 \frac{\partial M}{\partial x} ,\qquad(94)$$

where

$I_{1,2}$ = current in coil 1,2
$N_{1,2}$ = number of turns in coil 1,2
M = mutual inductance between one turn of coil 1 and one turn of coil 2.

Equation (94) can be derived from basic magnetic energy considerations for two interacting current-carrying coils. Appendix D contains the derivation.

The general expression for M can be shown[46] to be as follows:

$$M(x) = \mu_0 (a_1 a_2)^{0.5} \left[\left(\frac{2}{k} - k \right) K - \frac{2}{k} E \right] ,\qquad(95)$$

where

c = speed of light
$a_{1,2}$ = radius of coil 1,2
k^2 = $4 \times a_1 \times a_2 / [x^2 + (a_1 + a_2)^2]$
x = distance between the two coils
μ_0 = permeability of free space.

The quantities K and E are, respectively, elliptical integrals of the first and second kind and are given by:

$$K(k) = \int_{\varphi=0}^{\varphi=\frac{\pi}{2}} \frac{d\varphi}{\left(1 - k^2 \sin^2 \varphi\right)^{0.5}} \tag{96}$$

and

$$E(k) = \int_{\varphi=0}^{\varphi=\frac{\pi}{2}} d\varphi \left(1 - k^2 \sin^2 \varphi\right)^{0.5} . \tag{97}$$

Hence, for the case under consideration here, the net force acting on the bucket coil due to the four drive coils is given by

$$F = N_D N_B I_D I_B \frac{\partial}{\partial x} \left(-M_{n+2} + M_{n+1} + M_n - M_{n-1}\right) , \tag{98}$$

where

I_D = current in drive coil
I_B = current in bucket coil
N_D = number of turns per drive coil
N_B = number of turns per bucket coil
$M_i(x)$ = mutual inductance between one turn of the bucket coil and one turn of drive coil i ($i = n-1, \ldots, n+2$).

Note the minus signs in front of the $n+2$d and $n-1$st coils, which provide retarding forces, and the plus signs in front of the nth and $n+1$st coils, which provide motive forces.

The mutual inductances are given by

$$M_i(x) = \mu_0 (a_D a_B)^{0.5} \left[\left(\frac{2}{k_i} - k_i\right) K(k_i) - \frac{2}{k_i} E(k_i) \right] , \tag{99}$$

where

a_D = radius of drive coil
a_B = radius of bucket coil
$k_{n-1}^2 = 4 \times a_D \times a_B / [(S+x)^2 + (a_D + a_B)^2]$
$k_n^2 = 4 \times a_D \times a_B / [(x + (a_D + a_B)^2]$
$k_{n+1}^2 = 4 \times a_D \times a_B / [(S-x)^2 + (a_D + a_B)^2]$
$k_{n+2}^2 = 4 \times a_D \times a_B / [(\{2 \times S\} - x)^2 + (a_D + a_B)^2]$.

To facilitate calculational procedures, the two elliptical integrals, $K(k)$ and $E(k)$, have both been curve fitted. This was done using raw data[48] accurate to the fourth decimal place. Both curves were fit to sixth-order polynomial equations. The resulting curves are shown in figure 116 (denoted by Poly($K(m)$) and Poly($E(m)$), respectively), superimposed upon the raw data. Note that the independent variable in the graph is denoted by m, which in terms of the quantities given above, is equal to k^2. The curve fit equations are displayed in equations (100) and (101).

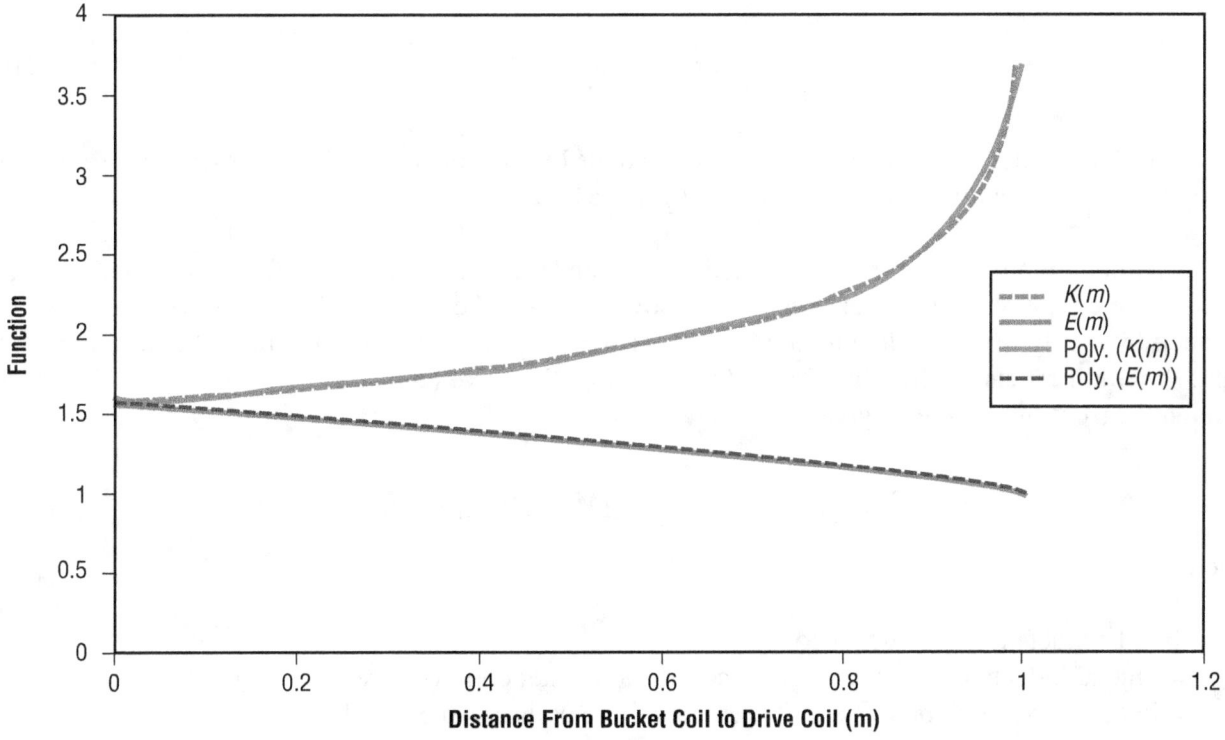

Figure 116. Elliptical function curve fits—$K(m)$ and $E(m)$ versus m.

Replacing m with k^2, the equations are as follows:

$$K(k^2) = 86.7180(k^2)^6 - 226.3862(k^2)^5 + 223.7493(k^2)^4 + 102.6237(k^2)^3$$
$$+ 22.0258(k^2)^2 - 1.4009(k^2) + 1.6024 \qquad (100)$$

$$E(k^2) = -1.9341(k^2)^6 + 4.9415(k^2)^5 - 4.8625(k^2)^4 + 2.1863(k^2)^3$$
$$- 0.5435(k^2)^2 - 0.3540(k^2) + 1.5701 \; . \qquad (101)$$

C.2 Drive and Bucket Coil Currents

Although the force equation is fairly straightforward, the calculations are complicated somewhat by the variations in both drive and bucket coil currents. The drive current can be written as

$$I_D = I_{D0} \sin\frac{\pi x}{S} \; , \qquad (102)$$

where x is the distance between the nth drive coil and the bucket coil and I_{D0} is the maximum current in the drive coil. In what follows, it is assumed that $(I_B/I_D) < 1$.

Although this relatively simple functional dependence can be assumed for I_D, the bucket current (I_B) is more complex. As the bucket carries no power source, the current will vary according to the flux supplied by the drive coils. The guiding principle is that the net flux through a bucket coil remains constant. When it first enters the mass driver, before any of the drive coils are energized, the total flux (Φ_1) through the front bucket coil is given by

$$\Phi_1 = N_B L I_0 - N_B M_{12} I_0 + N_B M_{13} I_0 - N_B M_{14} I_0 \; , \qquad (103)$$

where

L = self-inductance of a bucket coil
M_{1j} = mutual inductance between bucket coil 1 and bucket coil j (where $j = 2,3,4$)
I_0 = initial current in a bucket coil (before any drive coils are energized).

The self-inductance of a bucket coil is given by:[49]

$$L = \mu_0 a_B \left(\ln\frac{8a_B}{r_o} - \frac{7}{4} \right), \qquad (104)$$

where

r_o = conductor wire radius.

The mutual inductances between bucket coils are given by

$$M_{ij}(x) = \mu_0 a_B \left[\left(\frac{2}{k_{ij}} - k_{ji} \right) K(k_{ij}) - \frac{2}{k_{ij}} E(k_{ij}) \right], \tag{105}$$

where

$k_{ii+1}{}^2 = k_{ii-1}{}^2 = 4 \times a_B{}^2/[S^2+(2 \times a_B{}^2)]$
$k_{ii+2}{}^2 = k_{ii-2}{}^2 = 4 \times a_B{}^2/[(2 \times S^2)+(2 \times a_B{}^2)]$
$k_{ii+3}{}^2 = k_{ii-3}{}^2 = 4 \times a_B{}^2/[(3 \times S^2)+(2 \times a_B{}^2)]$

and the K and E functions are the same as those given earlier.

Note that the signs of alternate mutual inductances are reversed to account for the current flow directions in alternate bucket coils, as shown in figure 117.

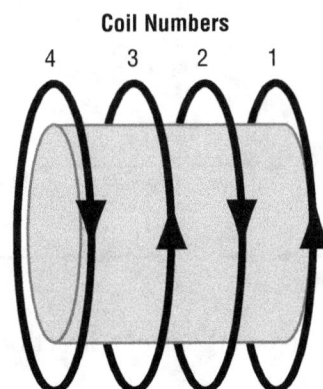

Alternating Current Flow Directions in Bucket Coils

Figure 117. Bucket coil current directions.

The expression for total flux can be simplified to

$$\Phi_1 = N_B L_1 I_0, \tag{106}$$

where L_1 is the total inductance of bucket coil 1, given by

$$L_1 = L + M_{12} + M_{13} + M_{14} \ . \tag{107}$$

Inspection of figure 117 shows immediately that the four coils will not all have the same total inductance; i.e., L_1 through L_4 are not all identical. The more centrally located coils will have higher values than those on the ends. Coils 1 and 4, the two end coils, will both have the same inductance as will coils 2 and 3, the middle coils. In other words, $L_1 = L_4$ and $L_2 = L_3$, but $L_1 \neq L_2$. Inspection shows that the differences are not large and that, typically, all four coils are within ≈5 percent of the average induction. To avoid having to develop separate induction models for the end and central coils for the purpose of this model, the average value is used.

When the bucket undergoes acceleration (fig. 118), the total flux through the coil is still Φ_1, but this is now given in terms of the drive coil current and various mutual inductances between drive and bucket coils by

$$\Phi_1 = N_B L_1 I_B + N_D I_D (M_{n+3} - M_{n+2} + M_{n+1} - M_n + M_{n-1}) \ . \tag{108}$$

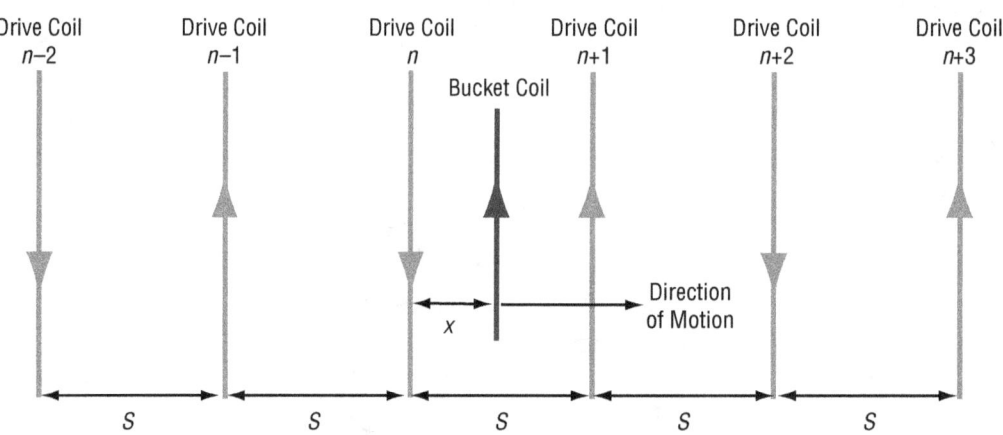

Figure 118. Bucket and drive coil current directions during acceleration.

Note the signs before each of the mutual inductances. The drive coils that are energized in the same sense as the bucket coil will add to the flux. Those energized in the opposite sense will subtract from it. Unlike the motive force equation, which only took into account the $n-1$, n, $n+1$, and $n+2$ drive coils, the effect of the $n+3$ coil is also included. This is because it adds to the flux through the bucket coil and hence diminishes its current. The lower current means that the motive force is reduced. This is in keeping with the conservative intent throughout this analysis. The $n-2$ drive coil is not included because it serves to increase the bucket coil current.

Eliminating Φ_1 gives

$$I_B = I_0 - N_D I_D \frac{M_{n+3} - M_{n+2} + M_{n+1} - M_n + M_{n-1}}{N_B L_1} \tag{109}$$

as the bucket coil current. As explained earlier, at any instant, all four-bucket coils have current of the same magnitude, but with flow direction alternating between successive coils.

C.3 Analysis of Bucket Kinetic Energy and Acceleration

The total mechanical work done on the bucket coil as it moves from the nth to the $n+1$st drive coil; i.e., from $x=0$ to $x=S$, is given by

$$\Delta E = \int_{x=0}^{x=S} F dx \ . \tag{110}$$

Although there are now expressions for I_D, I_B, and the various mutual inductances, analytical integration of the force equation is impractical. A numerical integration, using the following simple difference equation, is conducted instead:

$$\delta KE = F \delta x = N_D N_B I_D I_B \delta \left(-M_{n+2} + M_{n+1} + M_n - M_{n-1} \right) , \tag{111}$$

where δKE is the incremental increase in bucket kinetic energy between x and $x + \delta x$. This can be rewritten as

$$\delta KE(x \to x + \delta x) = N_D N_B I_D(x) I_B(x) \left\{ -\left[M_{n+2}(x+\delta x) - M_{n+2}(x) \right] \right.$$

$$+ \left[M_{n+1}(x+\delta x) - M_{n+1}(x) \right] + \left[M_n(x+\delta x) - M_n(x) \right]$$

$$\left. - \left[M_{n-1}(x+\delta x) - M_{n-1}(x) \right] \right\} , \tag{112}$$

where

$$I_D(x) = I_{D0} \sin \frac{\pi x}{S} \tag{113}$$

and

$$I_B(x) = I_0 - N_D I_D \frac{M_{n+3}(x) - M_{n+2}(x) + M_{n+1}(x) - M_n(x) + M_{n-1}(x)}{N_B L_1} \ . \tag{114}$$

These three equations can be integrated in a stepwise fashion to determine the total kinetic energy increase that the bucket experiences due to a single one of its coils as it traverses from $x=0$ to $x=S$. This is then multiplied by 4 to account for all the bucket coils, thus giving the total increase in bucket kinetic energy, denoted by ΔKE.

Note that the model is completely independent of the bucket speed. This means that ΔKE is independent of the bucket speed. For a design such as this, with equally spaced drive (and bucket) coils, this means that the same increase in kinetic energy is experienced between each pair of drive coils. Thus, one can write the final bucket speed (v), at the end of the accelerating portion of the mass driver, as

$$v \approx \sqrt{2 N_{DC} \frac{\Delta KE}{M_B}} \quad , \tag{115}$$

where N_{DC} is the total number of drive coils. The approximate equality symbol is used in the equation because the model employed here clearly does not apply at the very start of the mass driver, when there are no drive coils behind the bucket. Similarly, it does not apply at the very end of the acceleration portion when there are no more accelerating coils ahead of the bucket. However, provided there is a sufficiently large number of drive coils, say at least 20, the above expression should suffice.

Note that although the bucket acceleration is not constant as it moves between two adjacent drive coils, the force profile climbs from zero at $x=0$, to maximize at approximately $x=S/2$, and then subsequently declines. The average acceleration between any two adjacent drive coils is the same all the way along the coil gun if one neglects end effects.

This can be seen as follows:

$$\frac{dv}{dt} = \frac{dx}{dt}\frac{dv}{dx} = v\frac{dv}{dx} = \frac{d(v^2/2)}{dx} \quad , \tag{116}$$

where x is the distance along the mass driver traveled by the bucket. If one denotes the mean acceleration between two drive coils by $\langle dv/dt \rangle$, then between the ith and $i+1$th drive coils,

$$\left\langle \frac{dv}{dt} \right\rangle = \frac{1}{2}\frac{v_{i+1}^2 - v_i^2}{S} \quad , \tag{117}$$

and thus

$$\left\langle \frac{dv}{dt} \right\rangle = \frac{1}{2}\frac{\Delta KE / M_B}{S} \quad , \tag{118}$$

which is a constant for all i. Hence, the average acceleration can be treated as being constant.

C.4 Drive Coils

Drive coil current has been expressed as a function of the distance (x) traveled by the bucket between two drive coils:

$$I_D(x) = I_{D0} \sin \frac{\pi x}{S} \:. \tag{119}$$

In practice, of course, the current must be expressed as a function of time.

If one considers a single drive coil, the ideal current versus time profile would be as depicted in figure 119. By following a square profile, the drive coil is always at the maximum value to attract or repel a bucket coil. In practice, a square profile will probably not be achievable and so, instead, assume a simple oscillating inductance-capacitance-resistance (LCR) circuit. If the effects of resistance are neglected, the ideal sinusoidal current versus time profile is as shown in figure 120.

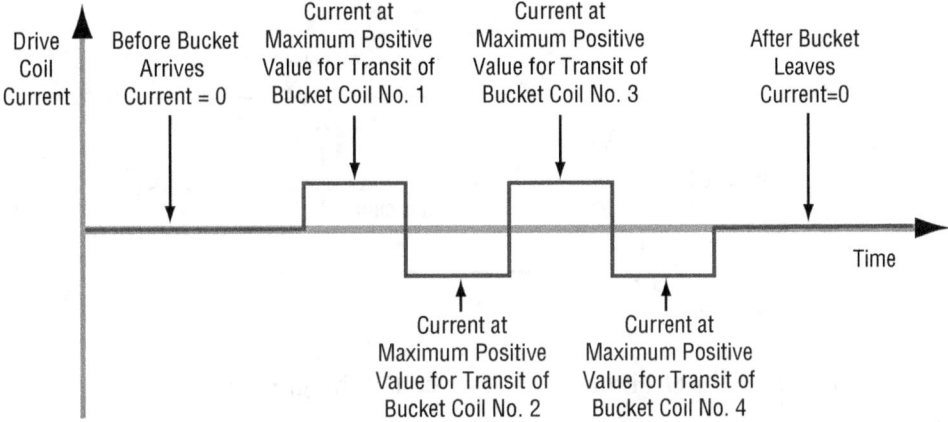

Figure 119. Ideal current versus time profile for a single drive coil.

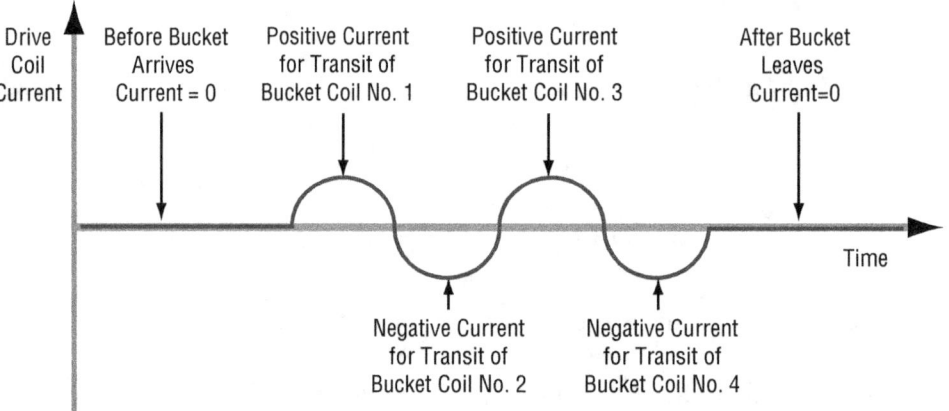

Figure 120. Sinusoidal current versus time profile for a single drive coil.

In theory, this type of current profile can be produced relatively easily by incorporating the drive coil into the type of circuit shown in figure 121.

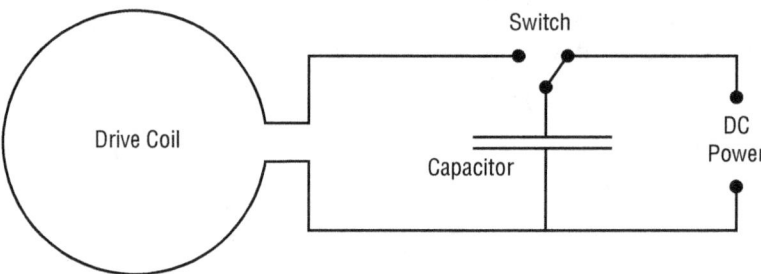

Figure 121. Oscillating drive circuit—charging capacitor.

The capacitor is charged from a DC power source (the switch is shown in fig. 121 in the charging position) and is then discharged to produce the required coil current, as shown in figure 122.

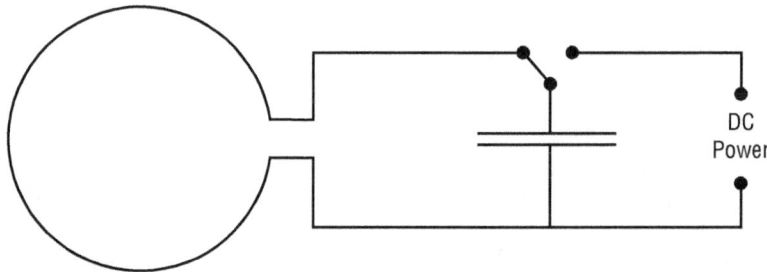

Figure 122. Oscillating drive circuit—discharging capacitor.

The LCR of the circuit are selected to give the required oscillation frequency, which will be different for each drive coil—earlier coils having a lower frequency than later coils. Resistance and other energy-loss mechanisms will result in a decaying current profile, as shown in figure 123.

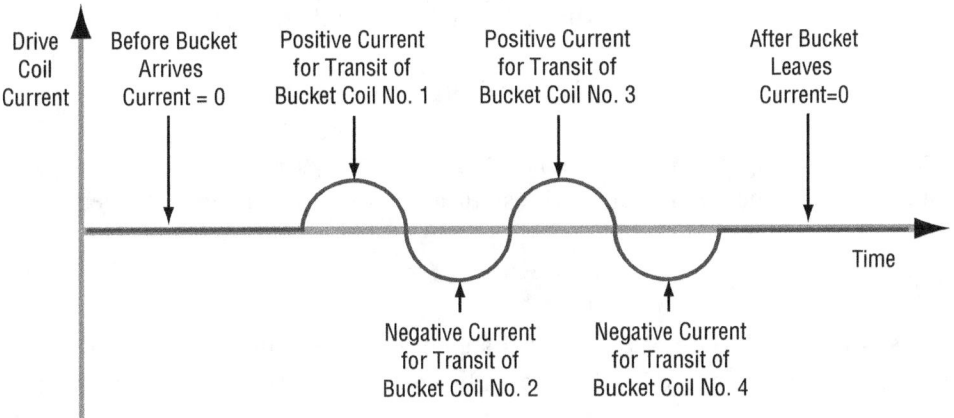

Figure 123. Real current versus time profile for a single drive coil.

Following the transit of one bucket, the circuit is switched back to the DC power source and the capacitor is charged again in preparation for the next bucket. Note that, although adequate for this simple model, the above drive coil circuit will need to be modified for practical use. This is because, unless $(I_B/I_D) \ll 1$, the drive current's sinusoidal profile for a simple circuit will be disrupted by the intrusion of bucket coil flux. In practice, to prevent this from happening, the drive coil electrical circuit will probably need to be fairly complex, with multiple capacitors being discharged in a carefully timed sequence as the bucket coil passes.

Drive coil circuit oscillation frequencies will vary along the length of the mass driver. As a coil's frequency is set by the time taken for the bucket to travel the standard intercoil distance (S), the earlier coils will have a lower frequency than the later coils. This means that each coil's circuit must be designed to produce the frequency (v) that is appropriate for the bucket's speed when it passes the coil:

$$v = \frac{v_i}{S} , \qquad (120)$$

where v_i is the mean speed of the bucket as it traverses the distance $2 \times S$, centered on the ith bucket.

This frequency-specific aspect of the drive coil design has two implications:

(1) A given drive coil interacts with each of the four bucket coils in succession. Its drive circuit should ideally be able to change its frequency slightly as each bucket coil passes. This is because the bucket speed will increase slightly as successive coils pass a given drive coil, in principle, necessitating a slightly higher frequency. The possibility of modifying drive circuit design to facilitate this should be investigated.

(2) The total mass of the loaded bucket must not vary significantly between "shots" of the mass driver. This means that the mass of expellant added to the bucket must be accurately metered. Use of a bucket whose empty mass is significantly greater than the expellant mass it carries would clearly help reduce this sensitivity. An alternate approach, provided that some variation in drive coil frequency were possible, would be to weigh the bucket by some means after it has been loaded with expellant. This might be done by vibrating the bucket and measuring either its amplitude or response frequency.

C.5 Braking Coils

Analysis of the braking process is conducted using exactly the same equations as for the acceleration process. The sole difference is that all of the stationary coil currents are reversed.

C.6 Bucket Design

The bucket is assumed to be of cylindrical shape. As shown in figure 124, it consists of an inner structural layer surrounded by a dewer that contains liquid nitrogen (LN$_2$) and the superconducting coils.

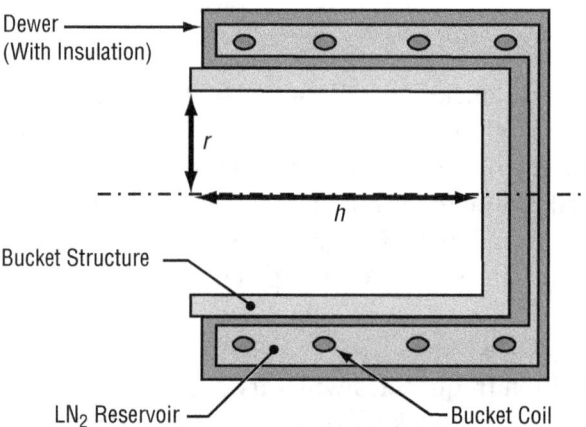

Figure 124. Bucket conceptual design.

The bucket shape is determined by the ratio of the internal height to the internal radius,

$$\eta = \frac{h}{r} \; ; \tag{121}$$

thus, the internal volume (V), available to hold expellant, is given by

$$V = \eta \pi r^3 \; . \tag{122}$$

The radius is given in terms of the volume and η by

$$r = \left(\frac{V}{\eta\pi}\right)^{\frac{1}{3}}. \tag{123}$$

The size of the LN_2 reservoir is dictated by the total heat dissipated from the bucket coils during one acceleration-deceleration cycle. The LN_2 will warm slightly during the acceleration-deceleration process due to the fact that bucket coil currents will vary due to inductive interaction with the stationary coils. If the total amount of heat generated is denoted by ΔQ and the allowable LN_2 temperature extremes are T_L and T_H, then, assuming that the thermal capacity of the coils may be neglected compared to that of the LN_2,

$$\Delta Q = (T_H - T_L) M_{LN_2} C_{LN_2}, \tag{124}$$

where M_{LN_2} is the mass of LN_2 and C_{LN_2} is specific heat.

Hence, the minimum mass of LN_2 required is

$$M_{LN_2} = \frac{\Delta Q}{(T_H - T_L) C_{LN_2}}. \tag{125}$$

The upper temperature limit is set by the critical temperature of the superconducting coils; i.e., the temperature above which their superconducting properties decline. The large number of so-called high-temperature superconductors now available, with critical temperatures at or above that of LN_2 at 1 atm, suggest that the normal LN_2 boiling point (77.4 K) as a good value for T_H.

To minimize the mass of the bucket, it is clearly desirable to minimize the mass of LN_2 that must be carried. However, there are limits to the value of T_L that are achievable. Some explanation of how the entire system works is appropriate here. Having completed its deceleration, the bucket LN_2 reservoir will be near its maximum temperature (T_H). The temperature will rise by an additional small amount as it makes its way via the return leg of the driver, back to the expellant loading hopper. This is simply because the ambient temperature on the asteroid surface is likely to be 50–100 K higher than that of the LN_2 reservoir. Note that the bucket design must ensure that the reservoir temperature never exceeds T_H.

When the bucket has returned to the hopper, it will be placed into thermal contact with a cold plate at a sufficiently low temperature and for a sufficiently long time period to restore its LN_2 reservoir to T_L. For reference, the entire bucket return concept, including reservoir cooling and expellant loading, is shown in figure 125.

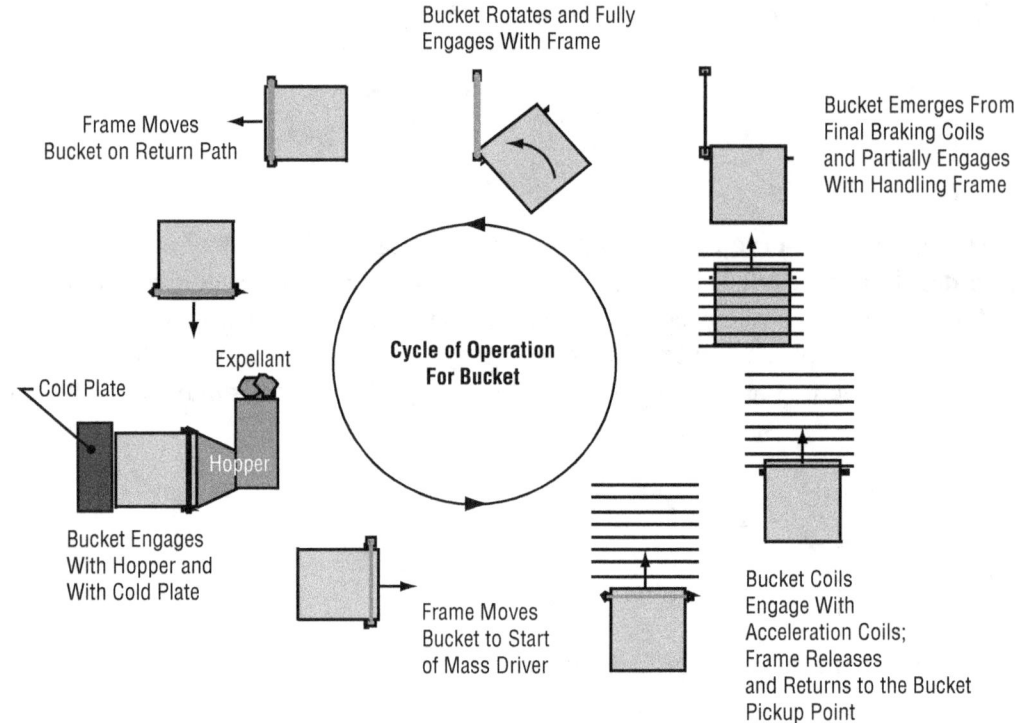

Figure 125. Bucket handling through a complete cycle.

The flow of heat from the bucket LN_2 reservoir to the cold plate can be written as

$$\frac{dQ}{dt} = Kr^2(T - T_{CP}) \;, \tag{126}$$

where K is a constant, r is the bucket external radius, which will probably determine the contact area with the cold plate, T is the instantaneous temperature of the bucket LN_2, and T_{CP} is the temperature of the cold plate. It is assumed that the cold plate is attached to a thermal reservoir and that its temperature may be taken as a constant.

For a given ΔQ, the above equation simply shows that the time which needs to be spent in contact with the cold plate decreases with increasing contact area; i.e., with r^2, and also decreases with decreasing cold plate temperature (T_{CP}).

This equation can be solved to give the time (τ) that the bucket must spend in contact with the cold plate in order to have its reservoir temperature lowered from T_H to T_L:

$$\tau = -\frac{H_C}{Kr^2} \ln\left(\frac{T_L - T_{CP}}{T_H - T_{CP}}\right), \tag{127}$$

where H_C is the total thermal capacity of bucket coils and LN_2.

The amount of heat dissipated in the bucket coils will, in some way, be proportional to the total time taken to accelerate and decelerate the bucket. Heat dissipation in the drive and braking coils is handled somewhat differently.

The current in a drive coil is given by

$$I_D(x) = I_{D0} \sin\left(\frac{\pi x}{S}\right). \tag{128}$$

If R is the resistance of the coil, then the total heat dissipated as the drive circuit oscillates through one cycle is given by

$$\text{Heat} = \int_{t=0}^{t=\frac{2\pi}{\omega}} I^2 R \, dt \; ; \tag{129}$$

hence,

$$\text{Heat} = \int_{t=0}^{t=\frac{2\pi}{\omega}} I_{D0}^2 R \sin^2(\omega t) \, dt . \tag{130}$$

Therefore,

$$\text{Heat} = \int_{x=0}^{x=2\pi} I_{D0}^2 \frac{R}{\omega} \sin^2 x \, dx \; ; \tag{131}$$

thus,

$$\text{Heat} = 2 I_{D0}^2 \frac{R}{\omega} \int_{x=0}^{x=\pi} \sin^2 x \, dx . \tag{132}$$

Hence,

$$\text{Heat} = 2 I_{D0}^2 \frac{R}{\omega} \frac{\pi}{2}$$

$$= \pi I_{D0}^2 \frac{R}{\omega} . \tag{133}$$

Now in practice, each drive coil will probably oscillate through about four complete cycles. Thus,

$$\text{Total heat dissipated} = 4\pi I_{D0}^2 \frac{R}{\omega} . \tag{134}$$

C.7 Interference Between Adjacent Drive Coils

As outlined earlier, it is anticipated that the drive coils could each be part of a simple inductance-capacitance circuit, tuned to the frequency appropriate to each coil's location along the length of the mass driver. This raises the question of how adjacent coils, which will have current flowing in opposite directions (fig. 126) will affect each other. This is a significant question given that, to avoid lengthening the bucket, which must be at least four coil spacings long, the coils must be relatively close to each other. The spacing would almost certainly need to be less than the stationary coil diameter.

Before considering the case of two interacting coils, it is useful to consider first a single isolated coil. For simplicity, the effects of resistance are neglected; they would be very minor anyway for superconducting coils.

The discharging capacitor circuit is as shown in figure 126.

As the capacitor begins to discharge, the increasing current induces an increasing magnetic flux (Φ) through the circuit (fig. 126). The increasing flux induces a back emf, acting against the direction of the increasing current:

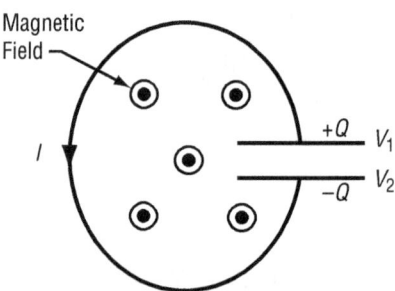

Figure 126. Discharge of isolated drive coil.

$$\text{Back emf} = \frac{d\Phi}{dt} . \tag{135}$$

From simple capacitor theory, one can see that

$$V_2 = V_1 - \frac{Q}{C} , \tag{136}$$

where C denotes the capacitance of the capacitor. From consideration of the back emf,

$$V_2 = V_1 - \frac{d\Phi}{dt} ; \tag{137}$$

thus,

$$\frac{d\Phi}{dt} = \frac{Q}{C} . \tag{138}$$

The capacitor charge and the discharge current are related via

$$I = -\frac{dQ}{dt} ; \tag{139}$$

so,

$$\frac{d^2\Phi}{dt^2} = -\frac{I}{C} . \tag{140}$$

The flux can be expressed in terms of the current and the circuit inductance (L) via

$$\Phi = LI . \tag{141}$$

To simplify the equations, the number of turns in the coil is assumed to have been taken into account in deriving the L value. This gives

$$\frac{d^2 I}{dt^2} + \frac{1}{L}CI = 0 , \tag{142}$$

which is the equation for simple harmonic motion and can be solved to give continuously oscillating solutions of the form

$$I = I_0 e^{i\omega t} , \tag{143}$$

where I_0 is the amplitude of the current oscillation and ω its angular frequency, given by

$$\omega = \left(\frac{1}{LC}\right)^{0.5}. \tag{144}$$

Now consider the case of two adjacent coils with instantaneous AC flow directions, as shown in figure 127.

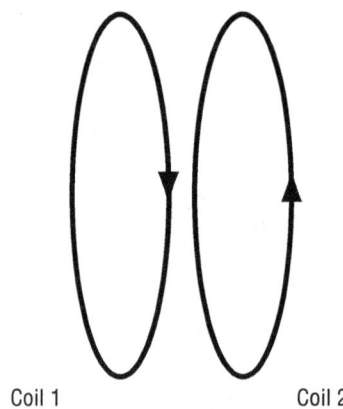

Figure 127. Current flow directions of two adjacent drive coils.

The interaction between these two coils can be described as follows. The total flux through coils 1 and 2 can be written as:

$$\Phi_1 = L_1 I_1 + M I_2 \tag{145}$$

and

$$\Phi_2 = L_2 I_2 + M I_1, \tag{146}$$

where

$\Phi_{1,2}$ = total flux through coils 1 and 2
$L_{1,2}$ = self-inductance of coils 1 and 2
$I_{1,2}$ = current in coils 1 and 2
M = mutual inductance of coils 1 and 2, which, as the coils are both fixed, remains a constant.

To simplify the equations, the number of turns in each coil is assumed to have been taken into account in the L and M values.

As for the case of the single coil, equation (147) can be written as

$$\frac{d^2\Phi_1}{dt^2} = -\frac{I_1}{C}, \tag{147}$$

from which Φ_1 can be eliminated to give

$$L_1 \frac{d^2 I_1}{dt^2} + M \frac{d^2 I_2}{dt^2} + \frac{I_1}{C} = 0. \tag{148}$$

Similarly, for circuit 2,

$$L_2 \frac{d^2 I_2}{dt^2} + M \frac{d^2 I_1}{dt^2} + \frac{I_2}{C} = 0. \tag{149}$$

Writing

$$I_1 = I_{10} e^{i\omega t} \tag{150}$$

and

$$I_2 = I_{20} e^{i\omega t}, \tag{151}$$

one can derive

$$I_1 \left[\frac{1}{C} - \omega^2 L \right] = M\omega^2 I_1 \tag{152}$$

and

$$I_2 \left[\frac{1}{C} - \omega^2 L \right] = M\omega^2 I_2. \tag{153}$$

These can be solved to give

$$\left[\frac{1}{C} - \omega^2 L \right]^2 = \left(M\omega^2 \right)^2; \tag{154}$$

hence,

$$\omega^2 = \frac{1}{C(L \pm M)} . \tag{155}$$

Substituting into one of the earlier equations relating the two currents gives

$$I_1 = \pm I_2 . \tag{156}$$

Hence, when the two currents flow in opposite directions, which is the situation here for adjacent drive coils, the minus sign applies and so

$$\omega^2 = \frac{1}{C(L-M)} . \tag{157}$$

This means that the frequency is shifted upward when compared to an isolated coil that had

$$\omega^2 = \frac{1}{CL} . \tag{158}$$

In addition to the frequency shift, the current will differ from that in an isolated coil. This can be seen by taking the expression

$$I_1 = I_{10} e^{i\omega t} \tag{159}$$

and integrating with respect to time to get an expression for the charge on the capacitor (Q_1):

$$Q_1 = \frac{I_{10}}{i\omega} e^{i\omega t} , \tag{160}$$

where I_{10} is the maximum current. Hence, the maximum charge on the capacitor (Q_{10}) is given by

$$Q_{10} = \frac{I_{10}}{i\omega} . \tag{161}$$

This means that, for a given maximum capacitor charge, the maximum current (I_{10}) is given by

$$I_{10} = i\omega Q_{10} . \tag{162}$$

Hence, for an isolated coil, the maximum current will be given by

$$I_{10} = \frac{Q_{10}}{(CL)^{0.5}}, \qquad (163)$$

whereas for an interacting coil,

$$I_{10} = \frac{Q_{10}}{(C(L-M))^{0.5}}. \qquad (164)$$

Hence, the discharge current is increased by a factor of

$$\frac{CL}{(C(L-M))^{0.5}} = \frac{1}{\left(1 - \frac{M}{CL}\right)^{0.5}} \qquad (165)$$

because of the interaction between the two coils. If the interaction is modeled between a coil and those with counter-running current on either side, the factor will be increased further to

$$\frac{1}{\left(1 - \frac{2M}{CL}\right)^{0.5}}. \qquad (166)$$

Note that, as the bucket proceeds along the mass driver, the drive coils are activated as they are required. Now from the above expression, it can be seen that the frequency of a coil will be somewhat lower when it is the endmost activated coil. As soon as it has energized neighbors on both sides, its frequency will increase. To avoid having this frequency shift take place while the bucket is very close, it may be necessary to activate each coil about a half period in advance and deactivate it about a half period after the ideal start and stop points.

C.8 Effect of Bucket Coil Motion on Stationary Coil Circuit Operation

A stationary coil can be represented as an L-C circuit as shown in figure 128.

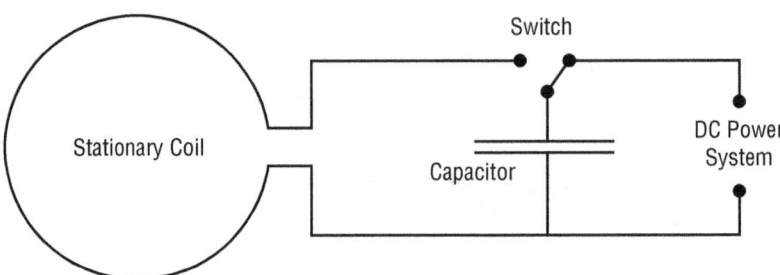

Figure 128. Simple model of a drive coil circuit.

For present purposes, it is not important to distinguish between the case of a drive coil and that of a braking coil. In the former case, the DC power system provides power to the stationary coil circuit; in the latter case, the DC power system accepts power from it.

Consider the situation where a bucket coil interacts with the stationary coil. As before,

$$\frac{d^2\Phi}{dt^2} = -\frac{I}{C} \tag{167}$$

for the stationary coil. In this case, the total flux is given by

$$\Phi = LI + MI' \;, \tag{168}$$

whereas before, L is the stationary coil self-inductance. The mutual inductance between stationary and bucket coils is denoted by M. As the bucket coil is in motion, this quantity does not remain constant. The current in the bucket coil circuit is denoted by I'. Note that, if the bucket coil current flows in the same direction as that of the stationary coil, $I' > 0$.

Hence,

$$\frac{I}{C} + \frac{d^2}{dt^2}(LI + MI') = 0 \;. \tag{169}$$

Now for the bucket coil, one can write

$$\Phi' = L'I' + MI \;; \tag{170}$$

thus,

$$I' = \frac{\Phi' - MI}{L'} \tag{171}$$

This enables one to eliminate I' from the main equation to give

$$\frac{I}{C} + \frac{d^2}{dt^2}\left(LI + M\frac{\Phi' - MI}{L'}\right) = 0, \tag{172}$$

which can be rewritten as

$$\frac{I}{C} + \frac{d^2}{dt^2}\left(\left(L - \frac{M^2}{L'}\right)I + M\frac{\Phi'}{L'}\right) = 0. \tag{173}$$

This can be expanded to give

$$\left[L - \frac{M^2}{L'}\right]\frac{d^2 I}{dt^2} + 2\frac{dI}{dt}\frac{d}{dt}\left[L - \frac{M^2}{L'}\right] + I\left[\frac{1}{C} + \frac{d^2}{dt^2}\left(L - \frac{M^2}{L'}\right)\right] = -\frac{d^2}{dt^2}\left(\frac{M\Phi'}{L'}\right). \tag{174}$$

Now L and L' are both constants. M depends upon the distance between the two coils, which is itself a function of time (t), and Φ' is a constant; hence, the equation can be rewritten as

$$f_1(t)\frac{d^2 I}{dt^2} + f_2(t)\frac{dI}{dt} + f_3(t)I = f_4(t), \tag{175}$$

where

$$f_1(t) = L - \frac{M^2}{L'}$$

$$f_2(t) = -\frac{2}{L'}\frac{d(M^2)}{dt}$$

$$f_3(t) = \frac{1}{C} - \frac{1}{L'}\frac{d^2(M^2)}{dt^2}$$

$$f_4(t) = -\frac{\Phi'}{L'}\frac{d^2 M}{dt^2} \ .$$

Now, sinusoidal solutions with changing amplitude can be expressed as

$$I = I_0 e^{(\alpha + i\omega)t} \ . \tag{176}$$

These solutions will only emerge from the complimentary equation corresponding to equation (169):

$$f_1(t)\frac{d^2 I}{dt^2} + f_2(t)\frac{dI}{dt} + f_3(t)I = 0 \ . \tag{177}$$

Solving this for solutions of the type

$$I = A e^{\lambda t} \ , \tag{178}$$

where λ is a constant, one obtains

$$f_1(t)\lambda^2 + f_2(t)\lambda + f_3(t) = 0 \ ; \tag{179}$$

hence,

$$\lambda = \frac{-f_2(t) \pm \sqrt{f_2(t)^2 - 4 f_1(t) f_3(t)}}{2 f_1(t)} \ . \tag{180}$$

Now in order that the oscillating solutions be obtained, it is necessary that the inequality

$$4 f_1(t) f_3(t) > f_2(t)^2 \tag{181}$$

be satisfied. If this is not met, then the stationary coil current will not oscillate and the mass driver, as designed here, will not function properly. For present purposes, simply assume that satisfying this inequality is a challenge for the detailed design.

If this condition is satisfied, then

$$\lambda = \frac{-\frac{f_2(t)}{2f_1(t)} \pm i\sqrt{4f_1(t)f_3(t) - f_2(t)^2}}{2f_1(t)} , \qquad (182)$$

which is of the form

$$\lambda = \alpha \pm i\omega , \qquad (183)$$

with

$$\alpha = -\frac{f_2(t)}{2f_3(t)} = \frac{\frac{2}{L'}\frac{d(M^2)}{dt}}{2\left(L - \frac{M^2}{L'}\right)} . \qquad (184)$$

Clearly, the sign of the above expression in a particular situation will determine whether the stationary coil current amplitude is increasing or decreasing. Before proceeding further, it is useful to consider the sign of the denominator in equation (184). First note that

$$\Phi = LI + M' \qquad (185)$$

and

$$\Phi' = L'I' + MI , \qquad (186)$$

which means that

$$\Phi - \frac{M\Phi'}{L'} = I\left(L - \frac{M^2}{L'}\right) ; \qquad (187)$$

hence,

$$L - \frac{M^2}{L'} = \frac{\Phi - \frac{M\Phi'}{L'}}{I} . \qquad (188)$$

Now, from the relative sizes of the drive and bucket coils as well as their relative currents, it can be said that

$$\Phi > \Phi' \text{ and } \frac{M}{L'} < 1 \ . \tag{189}$$

Hence, it follows that

$$L - \frac{M^2}{L'} > 0 \ . \tag{190}$$

Thus far, whether the stationary coil is a drive or a braking coil has not been specified. Also, it has not been stated whether the bucket coil is approaching or receding from the stationary coil.

C.9 Drive Coil

The following scenarios are for the bucket coil as it approaches and recedes from the drive coil:

- Bucket coil approaches drive coil: In this case, $dM/dt > 0$; hence, $d(M^2)/dt > 0$. This means that $\alpha > 0$, and the current amplitude increases.

- Bucket coil recedes from drive coil: In this case, $dM/dt < 0$; hence, $d(M^2)/dt < 0$. This means that $\alpha < 0$, and the current amplitude decreases.

To summarize, the drive coil current amplitude increases as the bucket coil approaches and decreases as it recedes. At first sight, this seems to imply a symmetry in the process, and might lead one to conclude that the drive coil's initial and final states are identical. Closer consideration shows that this is not the case. The magnitude of $d(M^2)/dt$ is set by the bucket speed. The faster the bucket travels, the greater will be $|d(M^2)/dt|$. Now, as the bucket is accelerating throughout, its speed of recession will be greater than its speed of approach. Consider two positions equidistant from and on either side of the drive coil, as shown in figure 129.

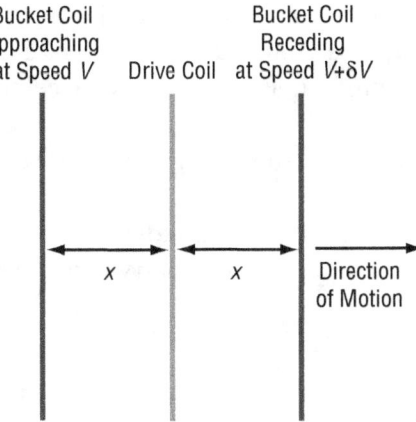

Figure 129. The effect of bucket acceleration.

At the first (left-hand) bucket coil position, $d(M^2)/dt$ is positive. At the second (right-hand) bucket coil position, $d(M^2)/dt$ is negative and of greater magnitude than it was at the first position. Thus, although these two positions are symmetric about the drive coil, the decrease in drive current amplitude at the second position exceeds the increase at the first position. This means that, after the bucket coil has passed, the drive coil current amplitude will be lower than it was initially. Hence, energy has been extracted from the drive coil circuit, despite the fact that resistance is neglected, simply by virtue of the bucket acceleration.

C.10 Braking Coil

The following scenarios are for the bucket coil as it approaches and recedes from the braking coil:

- Bucket coil approaches braking coil: In this case, $dM/dt > 0$; hence, $d(M^2)/dt > 0$. This means that $\alpha > 0$, and the current amplitude increases.

- Bucket coil recedes from braking coil: In this case, $dM/dt < 0$; hence, $d(M^2)/dt < 0$. This means that $\alpha < 0$, and the current amplitude decreases.

These results are identical to those for the drive coil. The difference is that, here the bucket is decelerating, so the amplitude increases while the bucket approaches and exceeds its decrease while the bucket recedes. Hence, energy is supplied to the braking coil circuit by virtue of the bucket deceleration.

Note that, although the bucket coil current rises and falls because of induction, the total magnetic flux through a bucket coil remains constant throughout the acceleration and deceleration processes. Hence, once again, neglecting resistance, the initial and final bucket coil currents are the same and there is no change in energy.

To summarize what is shown above, even with a traditional zero-resistance model, which for an isolated circuit just produces an undamped sinusoidal oscillation, energy is lost from a drive coil circuit and supplied to a braking coil circuit simply due to the bucket acceleration and deceleration.

C.11 Self-Induced Coil Stresses

A current-carrying coil loop will experience a radially outward force due to the interaction of its current with its self-generated magnetic field. The geometry is depicted in figure 130.

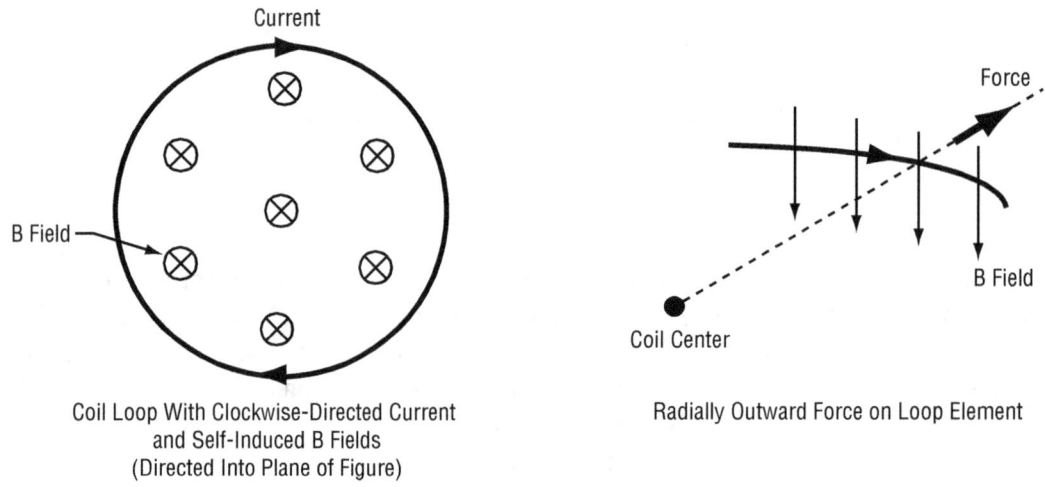

Figure 130. Self-induced magnetic field and resulting force.

The radial force on the entire coil (F) is best calculated by considering the change in energy associated with a slight increase in coil radius.

The total magnetic energy (U) of a system of N circuits is given by

$$U = \frac{1}{2}\sum_{i=1}^{N} I_i \Phi_i \ , \tag{191}$$

where I_i is the current through circuit i and Φ_i is the flux through circuit i.

In this case, each circuit is a turn of the coil; hence, all are identical. This means that

$$U = \frac{1}{2} N I \Phi \ , \tag{192}$$

where I and Φ are the current and flux through each circuit, respectively.

Now, the flux through one turn; i.e., one circuit, of the coil is due to the combined effects of the fields produced by each of the N turns. Hence,

$$\Phi = NLI ,\qquad(193)$$

where L is the self-inductance of a single turn; thus,

$$U = \frac{1}{2}LN^2I^2 .\qquad(194)$$

Therefore,

$$F = \frac{1}{2}I^2N^2\frac{dL}{dR} ,\qquad(195)$$

where R the radius of the coil.

The self-inductance of a single turn is given by equation (104):

$$L = \mu_0 R\left[\ln\left(\frac{8R}{r_0}\right) - \frac{7}{4}\right] ;\qquad(196)$$

hence,

$$\frac{dL}{dR} = \mu_0\left[\ln\left(\frac{8R}{r_0}\right) - \frac{3}{4}\right] .\qquad(197)$$

Thus, the total radial force on the entire coil is given by

$$F = \frac{1}{2}I^2N^2\mu_0\left[\ln\left(\frac{8R}{r_0}\right) - \frac{3}{4}\right] .\qquad(198)$$

The radial force per unit length of hoop per coil (f) is given by

$$f = \frac{F}{2\pi RN} ;\qquad(199)$$

thus,

$$f = \frac{1}{4\pi R} I^2 N \mu_0 \left[\ln\left(\frac{8R}{r_0}\right) - \frac{3}{4} \right] . \qquad (200)$$

The tension (T) within a loop is thus given by

$$T = Rf \; ; \qquad (201)$$

so,

$$T = \frac{1}{4\pi} I^2 N \mu_0 \left[\ln\left(\frac{8R}{r_0}\right) - \frac{3}{4} \right] . \qquad (202)$$

APPENDIX D—COIL FORCE MODEL

Consider an interacting set (N) of current-carrying circuits, each with inductance, capacitance, and resistance, and each subject to an externally imposed voltage. For circuit k, one can write

$$\left(V_k^{ext} + V_k\right) - R_k I_k - \frac{d\Phi_k}{dt} = 0 \;, \tag{203}$$

where

V_k^{ext} = voltage applied to circuit k from external sources
V_k = voltage across the capacitor of circuit k
R_k = resistance of circuit k
I_k = current in circuit k
Φ_k = total magnetic flux through circuit k, from both the circuit itself and all the other $N-1$ circuits.

The capacitance emf can be related to the current by

$$I_k = -C_k \frac{dV_k}{dt} \;, \tag{204}$$

where C_k is the total capacitance of circuit k.

The total available power in the entire system (W) is given by

$$W = \sum_k V_k^{ext} I_k - \sum_k R_k I_k^2 \;, \tag{205}$$

where the sum over all values of k is denoted by $\sum_k$.

Equation (203) can be used to eliminate V_k^{ext} from equation (205) to give

$$W = \sum_k I_k \frac{d\Phi_k}{dt} - \sum_k I_k V_k \;, \tag{206}$$

which can be rewritten, using equation (204) to give

$$W = \sum_k I_k \frac{d\Phi_k}{dt} + \frac{d}{dt}\sum_k \frac{1}{2}C_k V_k^2 \quad , \tag{207}$$

where it is assumed that none of the capacitances vary with time, even though the circuits may move relative to one another. In view of the ultimate application of this model, this assumption may seem questionable; however, it is only the fixed stationary circuits that have large capacitors built into their circuits. The bucket coils will have relatively low capacitances.

Noting that the total electrostatic energy (U_e) is given by

$$U_e = \frac{1}{2}\sum_k C_k V_k^2 \quad . \tag{208}$$

This gives

$$W = \sum_k I_k \frac{d\Phi_k}{dt} + \frac{dU_e}{dt} \quad . \tag{209}$$

Now, in general, the quantity W can be decomposed as follows:

$$W = \frac{d(U_e + U_m)}{dt} + W_{mech} \quad , \tag{210}$$

where the total magnetic energy, denoted by U_m and W_{mech}, represents the mechanical work done in moving the circuits about.

Combining equations (209) and (210) gives

$$\frac{dU_m}{dt} + W_{mech} = \sum_k I_k \frac{d\Phi_k}{dt} \quad . \tag{211}$$

Next, note that the total magnetic flux through a circuit is comprised of contributions due to its own inductance as well as that of each of the remaining circuits. Hence,

$$\Phi_k = \sum_i L_{ki} I_i \quad , \tag{212}$$

where L_{ki} denotes the mutual inductance on circuit k due to circuit i.

Hence, equation (211) can be written as

$$\frac{dU_m}{dt} + W_{mech} = \sum_k \sum_i I_k \frac{d(L_{ki}I_i)}{dt} \ . \tag{213}$$

Now, if one considers a restricted situation in which all the circuits are held securely and cannot engage in any relative motion, then it is clear that:

$$W_{mech} = 0 \text{ and } \frac{dL_{ki}}{dt} = 0 \tag{214}$$

for all values of k and i.

In this case,

$$\frac{dU_m}{dt} = \sum_k \sum_i I_k L_{ki} \frac{dI_i}{dt} \ , \tag{215}$$

from which it follows that

$$\frac{dU_m}{dt} = \frac{d}{dt} \sum_k \sum_i \frac{1}{2} I_k L_{ki} I_i \ . \tag{216}$$

This permits one to write the following expression for the magnetic energy. The expression is analogous to that for electrostatic energy:

$$U_m = \frac{1}{2} \sum_k \sum_i I_k L_{ki} I_i \ . \tag{217}$$

Now, consider the more general situation, where the circuits are permitted to engage in relative motion. Let the instantaneous location of all the circuits be defined by a set of parameters a_r, where $r=1, \ldots, s$.

Then, the work done in a movement (da_r) can be written as

$$\text{Work done} = F_r da_r \ ; \tag{218}$$

hence,

$$W_{mech} = \sum_{r=1}^{r=s} F_r \frac{da_r}{dt} \ . \tag{219}$$

But from equation (213), it is also known that

$$W_{mech} = \sum_k \sum_i I_k \frac{d(L_{ki} I_i)}{dt} - \frac{dU_m}{dt} \quad , \tag{220}$$

which can be rewritten using equation (217) as

$$W_{mech} = \sum_k \sum_i I_k \frac{d(L_{ki} I_i)}{dt} - \frac{d}{dt} \frac{1}{2} \sum_k \sum_i L_{ki} I_k I_i \quad ; \tag{221}$$

hence,

$$W_{mech} = \frac{1}{2} \sum_k \sum_i I_k \frac{d(L_{ki} I_i)}{dt} - \frac{1}{2} \sum_k \sum_i L_{ki} I_i \frac{dI_k}{dt} \quad . \tag{222}$$

This can be further simplified to

$$W_{mech} = \frac{1}{2} \sum_k \sum_i I_k I_i \frac{dL_{ki}}{dt} \quad . \tag{223}$$

Now, L_{ki} denotes the mutual inductance on circuit k due to circuit i. Its variation with time is due solely to the relative motion of the circuits; thus, one can write

$$\frac{dL_{ki}}{dt} = \sum_r \frac{dL_{ki}}{da_r} \frac{da_r}{dt} \quad , \tag{224}$$

which can be used together with equations (219) and (223) to give

$$\sum_r F_r \frac{da_r}{dt} = \frac{1}{2} \sum_k \sum_i \sum_r I_k I_i \frac{dL_{ki}}{da_r} \frac{da_r}{dt} \quad . \tag{225}$$

From this, it follows that

$$F_r = \frac{1}{2} \sum_k \sum_i I_k I_i \frac{dL_{ki}}{da_r} \quad . \tag{226}$$

Now, equation (226) is applied to the specific case of two circuits that are in relative motion:

$$F_r = \frac{1}{2}\left(I_1^2 \frac{dL_{11}}{da_r} + I_1 I_2 \frac{dL_{12}}{da_r} + I_2 I_1 \frac{dL_{21}}{da_r} + I_2^2 \frac{dL_{22}}{da_r}\right). \quad (227)$$

Now, in the case of interest here, with two circular coils in relative motion, there is no distortion of the coils, so the two self-inductance terms (L_{11} and L_{22}) are constant. Also, it can be shown from basic energy considerations that $L_{ki} = L_{ik}$ for all i and k. Hence,

$$F_r = I_1 I_2 \frac{dL_{12}}{da_r}. \quad (228)$$

For two coaxial coils, having N_1 and N_2 turns, respectively, and with motion along only the x axis, this can be rewritten as

$$F_x = N_1 N_2 I_1 I_2 \frac{dM}{dx}, \quad (229)$$

where M is the mutual induction between a single turn of each coil.

REFERENCES

1. Adams, R.B.; Alexander, R.A.; Chapman, J.M.; et al.: "Conceptual Design of In-Space Vehicles for Human Exploration of the Outer Planets," *NASA/TP–2003–212691*, Marshall Space Flight Center, MSFC, AL, November 2003.

2. French, B.M.: *Traces of Catastrophe*, Lunar and Planetary Institute, pp. 3, 14–15, 1998.

3. Adushkin, V.V.; and Nemchinov, I.V.: "Consequences of Impacts of Cosmic Bodies on the Surface of the Earth," *Hazards Due to Comets and Asteroids*, The University of Arizona Press, Tucson, pp. 721–778, 1994.

4. "The Barringer Meteorite Crater," http://www.barringercrater.com, The Cyrus Company, Cited December 12, 2003.

5. Rampino, M.R.; and Haggerty, B.M.: "Extraterrestrial Impacts and Mass Extinctions of Life," *Hazards Due to Comets and Asteroids*, The University of Arizona Press, Tucson, AZ, pp. 827–857, 1994.

6. Levy, D.: *Impact Jupiter: The Crash of Comet Shoemaker–Levy 9*, Plenum Press, New York, 1995.

7. "The Inner Solar System," Harvard-Smithsonian Minor Planet Center, http://cfa-www.harvard.edu/iau/Animations/Inner.gif, Cited March 2003.

8. Hills, J.G.; Nemchinov, I.V.; Popov, S.P.; and Teterev, A.V.: "Tsunami Generated By Small Asteroid Impacts," T. Gehrels (ed.), *Hazards Due to Comets and Asteroids*, The University of Arizona Press, Tucson, pp. 779–789, 1994.

9. Toon, O.B.; Zahnle, K.; Turco, R.P.; and Covey, C.: "Environmental Perturbations Caused by Asteroid Impacts," T. Gehrels (ed.), *Hazards Due to Comets and Asteroids*, The University of Arizona Press, Tucson, AZ, pp. 791–826, 1994.

10. Morrison, D.; Chapman, C.R.; and Slovic, P.: "The Impact Hazard," T. Gehrels (ed.), *Hazards Due to Comets and Asteroids*, The University of Arizona Press, Tucson, pp. 59–91, 1994.

11. Center for International Earth Science Information Network, Columbia University, http://www.ciesin.org Cited March 2003.

12. Lynn, E.: "INTROS: Integrated Rocket Sizing Program," Marshall Space Flight Center, AL, 2003.

13. Adams, R.; and Straw, A.: "Launch Vehicle Conceptual Design," Chapter 3 in *Space Launch and Transportation Systems*, pp. 79–214, to be published Spring 2004.

14. Venetoklis, P.; Gustafson, E.; Maise, G.; and Powell, J.: "Application of Nuclear Propulsion to NEO Interceptors," *Hazards Due to Comets and Asteroids*, The University of Arizona Press, Tucson, pp. 1089–1110, 1994.

15. Schmidt, G.R.; Bonometti, J.A.; and Irvine, C.A.: "Project Orion and Future Prospects for Nuclear Pulse Propulsion," *J. Propulsion and Power*, Vol. 18, pp. 497–504, May–June 2002.

16. Reynolds, T.W.: "Effective Specific Impulse of External Nuclear Pulse Propulsion Systems," NASA Technical Note, *NASA TN D–6984*, 1972.

17. Cotter, T.P.: "Rotating Cable Pusher for Pulsed-Propulsion Space Vehicle," Los Alamos Scientific Laboratory, *LA–4666–MS UC–33*, Propulsion Systems and Energy Conversion TID–4500, 1971.

18. Nance, J.C.: "Nuclear Pulse Propulsion," *IEEE Transactions on Nuclear Science*, February 1965.

19. Martin, A.R., and Bond, A.: "Nuclear Pulse Propulsion: A Historical Review of an Advanced Propulsion Concept," *J. British Interpl. Soc.*, Vol. 32, pp. 283–310, 1979.

20. Andrews, J.: and Andrews, D.: "The MagOrion—A Propulsion System for Human Exploration of the Outer Planets," American Institute of Physics, New York, pp. 1533–1537, 2000.

21. Solem, J.C.: "Medusa: Nuclear Explosive Propulsion for Interplanetary Travel," *J. British Interpl. Soc.*, Vol. 46, pp. 21–26, 1993.

22. Dyson, G.: *Project ORION: The True Story of the Atomic Spaceship*, Owl Books, 2002.

23. Bonometti, J.A.; Morton, P.J.; and Schmidt, G.R.: "External Pulsed Plasma Propulsion and Its Potential for the Near Future," American Institute of Physics, New York, pp. 1236–1241, 2000.

24. Bonometti, J.A.; and Morton, P.J.; "External Pulsed Plasma Propulsion (EPPP) Analysis Maturation," *AIAA–2000–3610*, 2000.

25. McInnes, C.: *Solar Sailing*, Praxis Publishing Ltd., Chichester, UK, pp. 33–109, 1999.

26. Wright, J.: *Space Sailing*, Taylor and Francis, Inc, pp. 225–239, 1992.

27. Serway, *Physics for Scientists and Engineers*, Saunders College Publishing, Philadelphia, 1986.

28. Ahrens, T.J.; and Harris, A.W.: "Deflection and Fragmentation of Near-Earth Asteroids," *Hazards Due to Comets and Asteroids*, The University of Arizona Press, Tucson, pp. 897–924, 1994.

29. "Dan's History," http://www.danshistory.com/nuke.shtml Cited April 14, 2003.

30. "Dan's History," http://danshistory.com/lgb.shtml Cited April 14, 2003.

31. Nelson, R.A.: "Low-Yield Earth Penetrating Nuclear Weapons," FSA Public Interest Report, p. 4, January/February 2001.

32. Melosh, H.J.; Nemchinov, I.V.; and Zetzer, Y.I.: "Non-Nuclear Strategies for Deflecting Comets and Asteroids," *Hazards Due to Comets and Asteroids*, The University of Arizona Press, Tucson, p. 1111–1132, 1994.

33. Harris, T.: "How Stuff Works," http://science.howstuffworks.com/e-bomb.htm Cited December 2002.

34. O'Neill, G.K.; and O'Leary, B.: "Space-Based Manufacturing From Nonterrestrial Materials," *Progress in Astronautics and Aeronautics*, published by the American Institute of Aeronautics and Astronautics, Vol. 57, 1977.

35. Ahrens, T.J.; and Harris, A.W.: "Deflection and Fragmentation of Near-Earth Asteroids," *Hazards Due to Comets and Asteroids*, The University of Arizona Press, Tucson, AZ, pp. 897–927, 1994.

36. Tedeschi, W.J.: "Mitigation of the NEO Impact Hazard Using Kinetic Energy," in Proc. Planetary Defense Workshop, Lawrence Livermore National Laboratory, Livermore, CA, pp. 313–323, May 22–26, 1995.

37. Bate, R.R.; Mueller, D.D.; and White, J.E.: *Fundamentals of Astrodynamics*, Dover Publications, Inc., New York, 1971.

38. Shoemaker, E.M.: "Asteroid and Comet Bombardment of the Earth," *Annual Rev. Earth and Planetary Sci.*, Vol. 11, pp. 461–494, May 1983.

39. Chapman, C.R.; and Morrison, D.: "Impacts on the Earth by Asteroids and Comets: Assessing the Hazard," *Nature*, Vol. 367, pp. 33–40, 6 January 1994.

40. Gold, R.E.: "SHIELD—A Comprehensive Earth Protection System: A Phase I Report to the NASA Institute for Advanced Concepts," Report No. SDO–10974, May 28, 1999.

41. Lewis, J.S.: *Comet and Asteroid Impact Hazards on a Populated Earth*, Academic Press, 1999.

42. Jeffers, S.V.; Manley, S.P.; Bailey, M.E.; and Asher, D.J.: "Near-Earth Object Velocity Distributions and Consequences for the Chicxulub Impactor," *Mon. Not. R. Astron. Soc.*, Vol. 327, p. 327, 2001.

43. Chesley, S.; Chodas, P.; Milani, A.; et al.: "Quantifying the Risk Posed by Potential Earth Impacts," *Icarus*, Vol. 159, pp. 423–432, 2002.

44. Ivezic, Z.; Tabachnik, S.; and Rafikov, R.: "Solar System Objects Observed in the Sloan Digital Sky Survey Commissioning Data," *Astron. J.*, Vol. 122, pp. 2749–2784, November 2001.

45. Brown, P.; Spalding, R.E.; ReVelle, D.O.; et al.: "The Flux of Small Near-Earth Objects Colliding with Earth," *Nature*, p. 142, 21 November 2002.

46. Reeves, E.I.: "Spacecraft Design and Sizing," Chapter 10 in *Space Mission Analysis and Design*, edited by James R. Wertz and Wiley J. Larson, Microcosm Press, El Segundo, CA, 1999.

47. NASA Glenn Research Center Web-page: http://spacescience.nasa.gov/osstech/solar.htm Cited December 2002.

48. Abramowitz, M.; and Stegun, I.A.: *Handbook of Mathematical Functions*, Dover Publications Inc., New York, NY, pp. 608–609, 1965.

49. Becker, R.: *Electromagnetic Fields and Interactions*, Dover Publications Inc., New York, NY, pp. 193–197, 1982.

www.ingramcontent.com/pod-product-compliance
Lightning Source LLC
Chambersburg PA
CBHW081723170526
45167CB00009B/3675